DRAFTING FOR TRADES & INDUSTRY

CIVIL

JOHN A. NELSON

DRAFTING FOR TRADES & INDUSTRY

CIVIL

JOHN A. NELSON

10 9 8 7 6 5 4 3

LIBRARY OF CONGRESS CATALOG CARD NUMBER: 77-91450
ISBN 0-8273-1844-8

Printed in the United States of America
Published simultaneously in Canada by
Delmar Publishers, A Division of
Van Nostrand Reinhold, Ltd.

DELMAR PUBLISHERS INC. • ALBANY, NEW YORK 12205

PREFACE

The size, shape, and contour of the earth is as important to our total environment as is a similar description of manufactured and constructed objects. Maps and charts are the documents created by cartographers and surveyors to communicate their measurements of the earth's features. Map drafting, or cartography, is the branch of drafting that collects data from surveyors and records it in the form of maps and charts.

Drafting for Trades and Industry – Civil is designed to develop the student's technical skills in mapmaking. The text explains the different types of maps and surveys and how to gather surveying information. The student will learn how to use this information to draw accurate maps, particularly contour maps.

While traditional texts present new material in narrative form, *Drafting for Trades and Industry – Civil* involves the student with extensive hands-on experience that applies drafting theories and develops skills. However, this is not simply a workbook. Each topic is developed through a progression of practice exercises that are augmented with explanations and suggestions to help the student overcome areas of difficulty.

In addition to the practice exercises, each unit includes a pretest and unit review. The pretest allows both the student and the instructor to determine where special attention is required. The unit review gives the instructor a means of evaluating the student's performance. Answers to all practice exercises are included within the units. Answers to pretests and unit reviews appear in a separate Instructor's Guide.

By providing the answers to practice exercises in the student's edition, this drafting program may be used in the traditional classroom or in an independent study program. Students using this book may advance at their own speed. The teacher is present to give guidance, explanations, and demonstrations when needed by the student. In this way, each individual can practice drafting techniques until they are mastered. The combination of instructor, proper instructional materials, and practice is the ideal way to develop a skill.

Drafting for Trades and Industry – Civil is designed for the student who has mastered basic drawing techniques. *Drafting for Trades and Industry – Basic Skills* provides this experience. Once basic drawing techniques are mastered, the student is ready to specialize in a particular field of drafting.

There are four areas on concentration in the *Drafting for Trades and Industry* series:

- *Drafting for Trades and Industry – Architectural*
- *Drafting for Trades and Industry – Civil*
- *Drafting for Trades and Industry – Technical Illustration*
- *Drafting for Trades and Industry – Mechanical and Electronic*

For a comprehensive overview of drafting, the student should complete all the books in the series. However, the student may be ready to seek employment as a civil drafter after completing *Basic Skills* and the material in the *Civil* concentration.

ABOUT THE AUTHOR

John A. Nelson is an experienced drafting instructor. He has spent 11 years working in design and drafting departments in industry and has 14 years experience as a drafting instructor. He has a degree from the College of San Mateo in California and has done post-graduate work at Foothill College in California and the University of Tennessee.

CONTENTS

INTRODUCTION

Drafting for Trades and Industry – Civil covers technical skills in mapmaking. Each topic is developed through a series of practice exercises which implement the information presented in the unit. There are no lectures. All material is illustrated and explained in the unit. The teacher provides additional guidance, explanations, and demonstrations when needed. In this way, students are free to work at their own pace and practice the basic skills until they are mastered.

Each student is expected to advance as rapidly as possible. The student should be familiar with the drafting reference books available in the classroom and library. These books provide alternate methods of drawing and additional exercises for the student who is having difficulty in a particular area of drafting. Be sure each point is fully understood. If something is unclear, ask the instructor.

UNIT INSTRUCTIONS

The *objective* explains the purpose of the unit. It outlines what information and skills the student will learn as a result of completing the exercises contained in the unit.

Most units begin with a *pretest* covering the material contained in the unit. Those who pass the pretest can go directly to the next unit. Those who are unfamiliar with the material or fail the pretest move through the unit and complete all the practice exercises.

Related terms are designed to teach the language of drafting. They are completed in the spaces provided after the information is read. Related terms should be thought of as study notes.

Each unit contains *information* which must be read before actual practice in drafting skills can take place. In some cases, information contains background material, while others contain instructions on how to perform particular skills or solve specific problems.

Information is followed by *practice exercises.* After completing each exercise, the work should be compared to the answer provided in the unit *before proceeding* to the next exercise or topic. Any questions should be discussed with the instructor before going on.

The *unit review* consists of an examination covering everything contained in the unit. It is taken after all work in the unit is completed. Read the instructions carefully, as many reviews have time limits within which the work must be done. If the unit review is successfully completed within the time limit, the student proceeds to the next unit of study.

The *flow chart* provides a graphic illustration of how the student progresses through the unit. By following the arrows, one can quickly see the procedure to follow to successfully complete each unit.

At the end of each unit review, there is room to record two things. First, the instructor may initial approval for the work completed in the unit. Second, the student may indicate that the unit's completion was recorded on the progress chart.

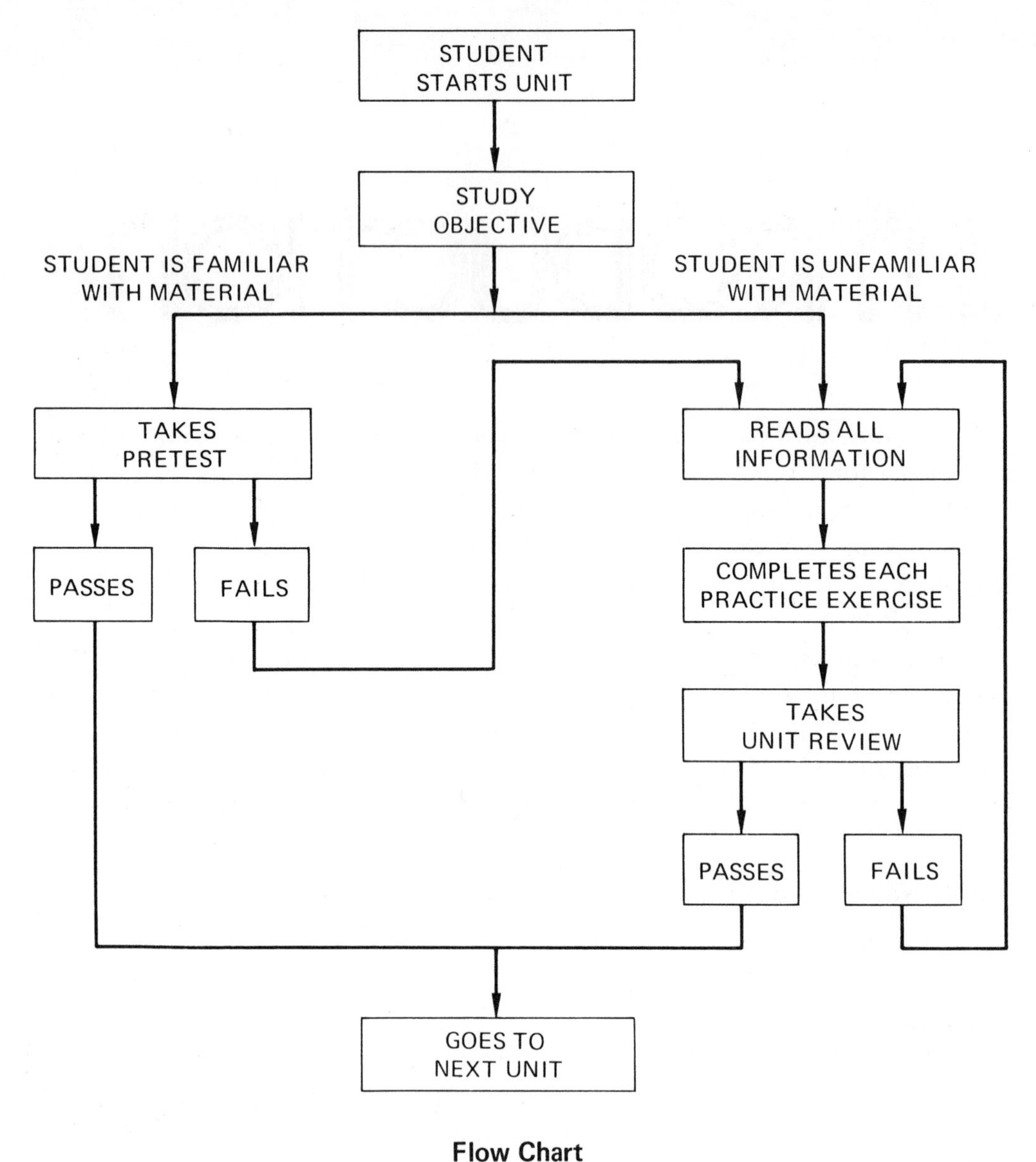

Flow Chart

PROGRESS CHART

The *progress chart* is a graph used to record the progress made by the student through the course. By using this chart, students know exactly where they are in the program and what units remain in order to complete all the work in the time allotted.

The numbers at the bottom of the graph indicate the number of school days. The unit titles are listed at the left. The "recommended progress" line indicates the average length of time the student should spend on each unit.

To record a unit's completion, place a dot where the unit title and the day the unit is completed meet. For example, *Unit 1 – Maps and Surveys* is completed on the 3rd day on the recommended progress line.

If the dot falls above the recommended progress line, then the student is ahead of schedule. If the dot falls below the line, the student is running behind schedule and should adjust the rate of progress if all the units are to be completed in time.

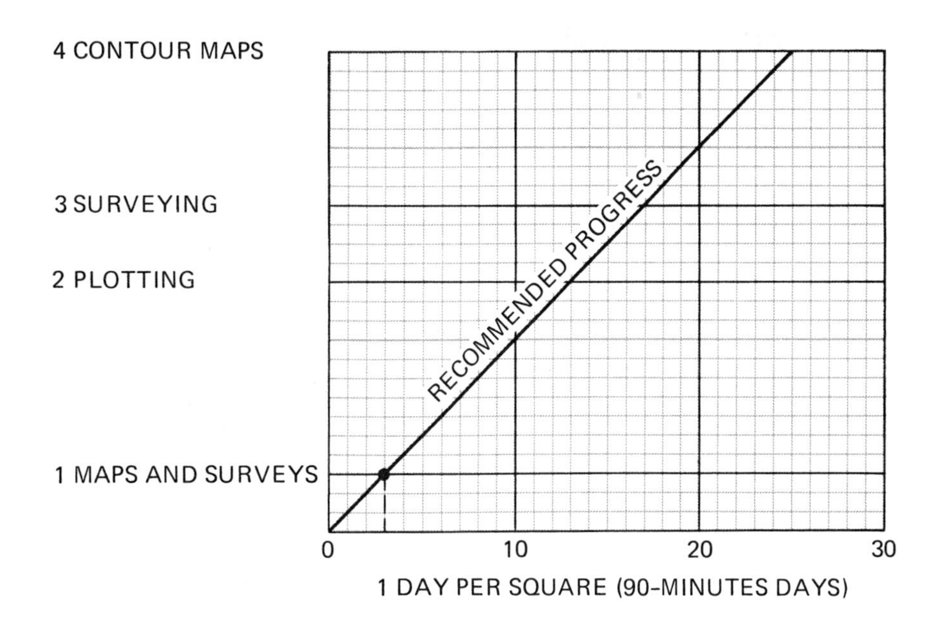

Progress Chart

APPLICATION CHART – Drafting for Trades and Industry

UNIT NO.	UNIT TITLE	Occupational Title: Tracer/Detailer	Architectural Drafter	Cartographer	Electronic Drafter	Technical Illustrator	Mechanical Drafter	Electro-Mechanical Drafter	Tool Designer	Drafting Teacher
Drafting for Trades and Industry–BASIC SKILLS										
1	Equipment	■	■	■	■	■	■	■	■	■
2	Lettering	■	■	■	■	■	■	■	■	■
3	Drawing Techniques	■	■	■	■	■	■	■	■	■
4	Geometric Construction	■	■	■	■	■	■	■	■	■
5	Multiview Drawings	■	■	■	■	■	■	■	■	■
6	Basic Isometrics	■	■	■	■	■	■	■	■	■
7	Section Views	■	■	■	■	■	■	■	■	■
8	Descriptive Geometry	■	■	■	■	■	■	■	■	■
9	Auxiliary Views	■	■	■	■	■	■	■	■	■
10	Developments	■	■	■	■	■	■	■	■	■
11	Basic Dimensioning	■	■	■	■	■	■	■	■	■
12	Careers in Drafting	■	■	■	■	■	■	■	■	■
Drafting for Trades and Industry–ARCHITECTURAL										
1	House Considerations		■							■
2	House Construction		■							■
3	Windows, Doors, Fireplaces, and Fixtures		■							■
4	Stair Layout		■							■
5	Structural Members and Loading		■							■
6	Working Drawings		■							■
Drafting for Trades and Industry–CIVIL										
1	Maps and Surveys		■	■						■
2	Plotting		■	■						■
3	Surveying		■	■						■
4	Contour Maps		■	■						■
Drafting for Trades and Industry–TECHNICAL ILLUSTRATION										
1	Mechanical Lettering		■			■				■
2	Advanced Isometrics		■			■				■
3	Perspective Drawing		■			■				■
4	Airbrush Techniques		■			■				■
Drafting for Trades and Industry–MECHANICAL AND ELECTRONIC										
1	Manufacturing Processes						■	■	■	■
2	Basic Welding						■	■	■	■
3	Fasteners						■	■	■	■
4	Precision Measurement						■	■	■	■
5	Springs						■	■	■	■
6	Cams						■	■	■	■
7	Advanced Dimensioning						■	■	■	■
8	Assembly and Detail Drawing						■	■	■	■
9	Electronic Components				■			■		■
10	Electronic Drawing				■			■		■
11	The Engineering Department				■		■	■	■	■

■ Required

The chart suggests the minimum number of units to cover for a particular field of interest. However, the final course content is the decision of the individual instructor.

7. Describe a boundary or land survey.

8. What is the point of beginning? Does it have to be on the property?

9. What is recommended at all property corners?

10. Where are all records of local land descriptions kept?

Before proceeding to the next unit:

_______ Instructor's approval

_______ Progress plotted

UNIT 2

PLOTTING

OBJECTIVE

The student will learn to plot traverses from field notes by using headings and distances.

PRETEST

90-minute time limit

Using the headings and distances listed, plot the enclosed traverse in the space provided on page 21. Be sure to include all inside angles. Use correct line weight and callouts.

Headings/Distances

A–B = S65°–00′E/745.0′
B–C = N18°–30′E/590.0′
C–D = S39°–00′E/700.0′
D–E = S60°–45′W/730.0′
E–F = S45°–45′E/890.0′
F–G = S75°–00′W/1085.0′
1) G–A = ________/______

2-8) ADD *ALL* INSIDE ANGLES
(7 answers req'd.)

RELATED TERMS

Give a brief definition of each term as progress is made through the unit.

Vernier ____________________

Deflection angle ____________________

Open traverse ____________________

Closed traverse ____________________

Directional arrowhead ____________________

PLOTTING DEFLECTION ANGLES

A *deflection angle* is an angle that is formed by a line to the right or left of the direction being traveled.

In figure 2-1, right is clockwise and left is counterclockwise. Follow the little footprints. Start at point A, work to point B, stand at point B looking ahead down the line of sight, turn to the right facing clockwise, and walk to point C. Stand at point C looking ahead down the line of sight, turn to the left facing counterclockwise to the next point. This is called *deflection angle plotting.*

It is suggested that the drafter line up the drafting machine on the line of sight with the arrow on the "head" set at 0 degree. Turn it either to the left or right and read the angle directly from the built-in protractor.

Pretest

SCALE 1″ = 200.0′

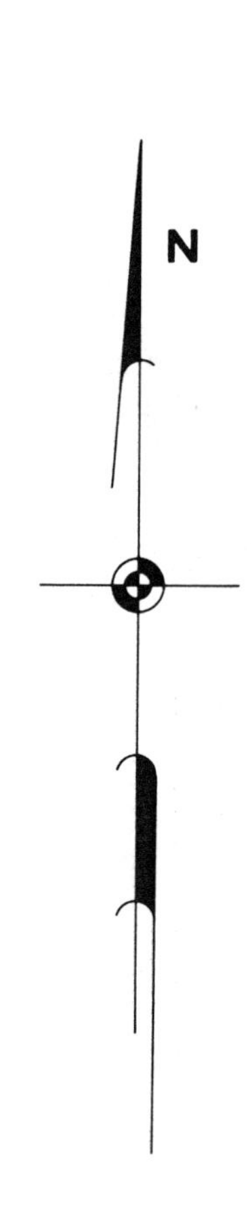

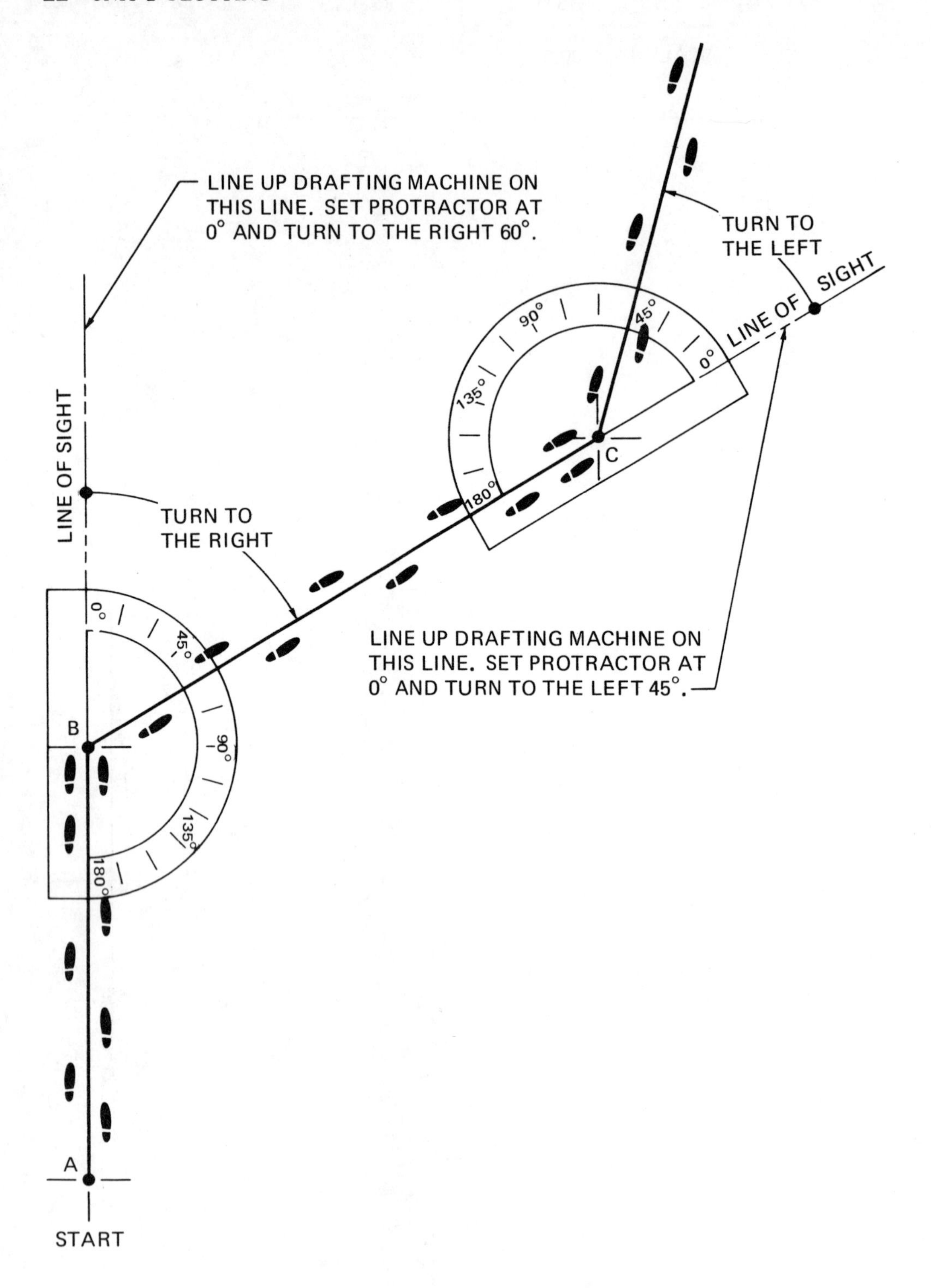

Fig. 2-1 Deflection angle plotting

TRAVERSE

A *traverse* is a series of connected lines of known lengths and directions. Connecting lines with known lengths and known directions form a traverse, figure 2-2. An *open traverse* is shown at the left, and a *closed traverse* is shown at the right. A closed traverse is easier to check as it must close upon itself and end where it started. An open traverse is more difficult to prove unless the drafter can sight back as illustrated by the dotted line, E to A, which would be just like a closed traverse.

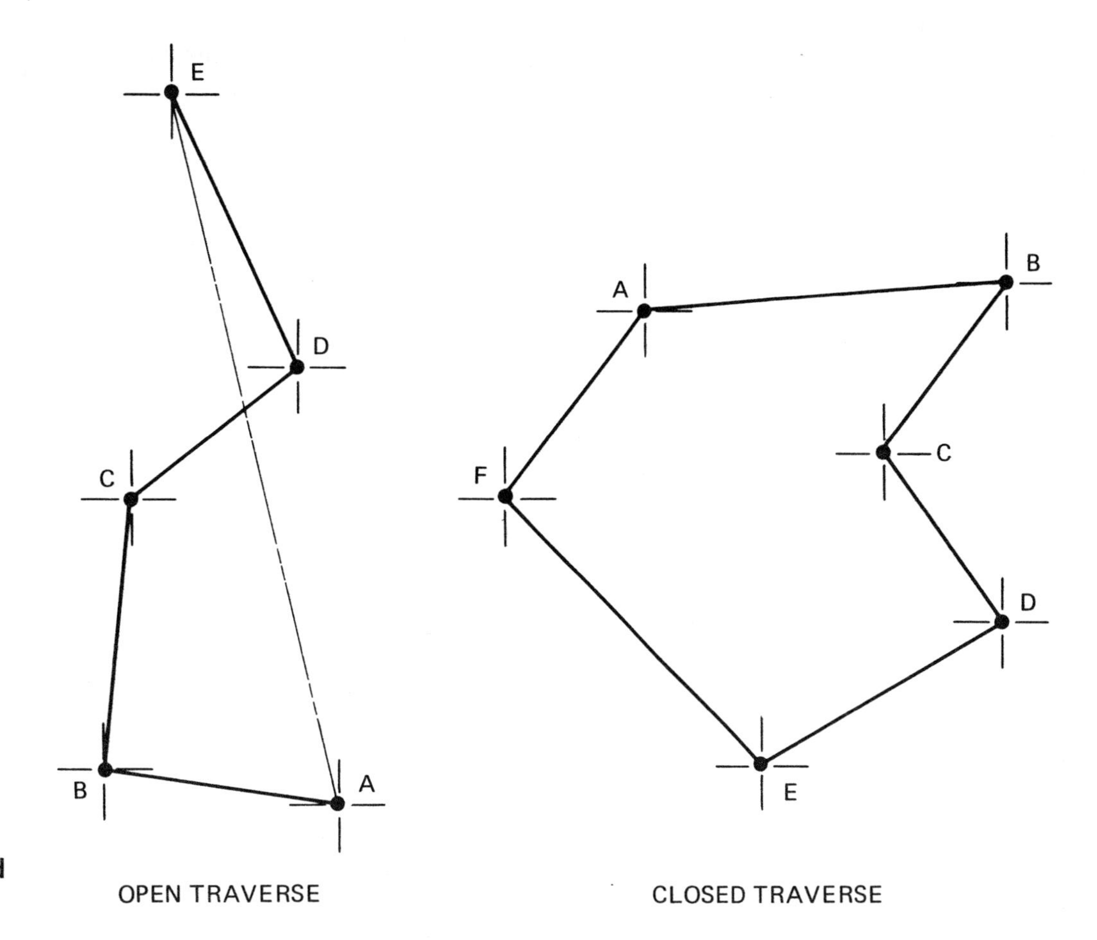

Fig. 2-2 Open and closed traverse

ARROWHEAD

Figure 2-3 shows a sample of a properly drawn *directional arrowhead.*

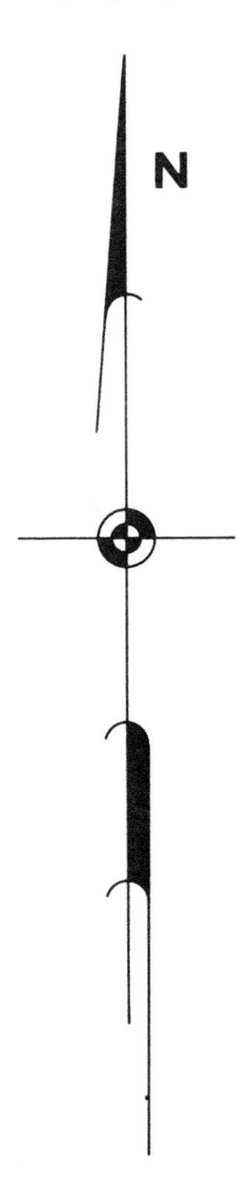

Fig. 2-3 Directional arrowhead

Always try to position the north arrow facing up on the drawing paper, figure 2-4. That way east will always be to the right, south at the bottom, and west to the left of the paper.

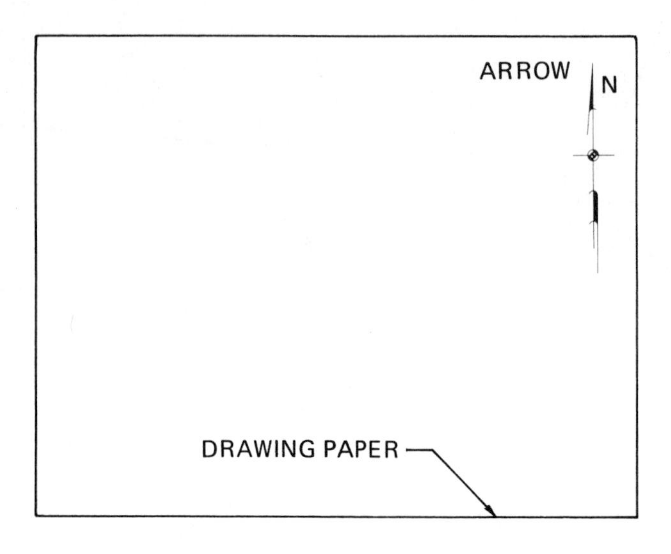

Fig. 2-4 Positioning the north arrow

DRAFTING STANDARDS

Line Weight. Use the same line contrast as in other drafting areas in order to make the drawing easier to understand.

Callouts. Each line must include the distance in feet and/or tenths and hundredths of a foot. Actual inches are not used. The bearing must also be included in the callout, figure 2-5.

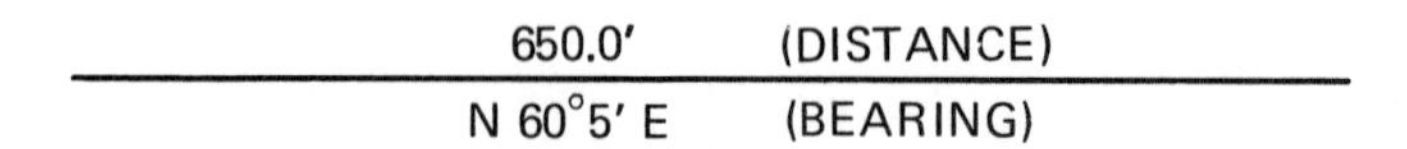

Fig. 2-5 Callout for lines

Corners. Neatly make each corner or turning point as illustrated in figure 2-6.

Fig. 2-6 Corner or turning point

A line from point (corner) A to point B is drawn exactly as shown in figure 2-7. All callouts must be parallel to the line, 1/8 inch high, and read from the bottom of the page.

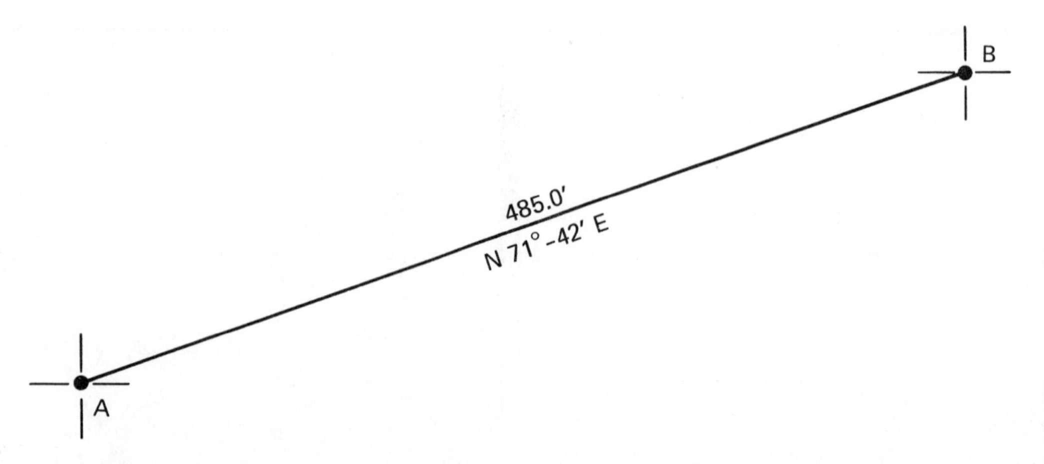

Fig. 2-7 Drawing a line from corner to corner

Practice Exercise 2-1

Using the land description given, carefully plot points A, B, C, D, E, and F. Draw a line from F to A to close the traverse. This line should be 380.0 feet long and have a S 30° W bearing. Line A–B is completed as an example. Label each line and point as illustrated in line A–B. Compare your work to the answer on page 42. Keep this drawing as it will be used in exercise 2-8.

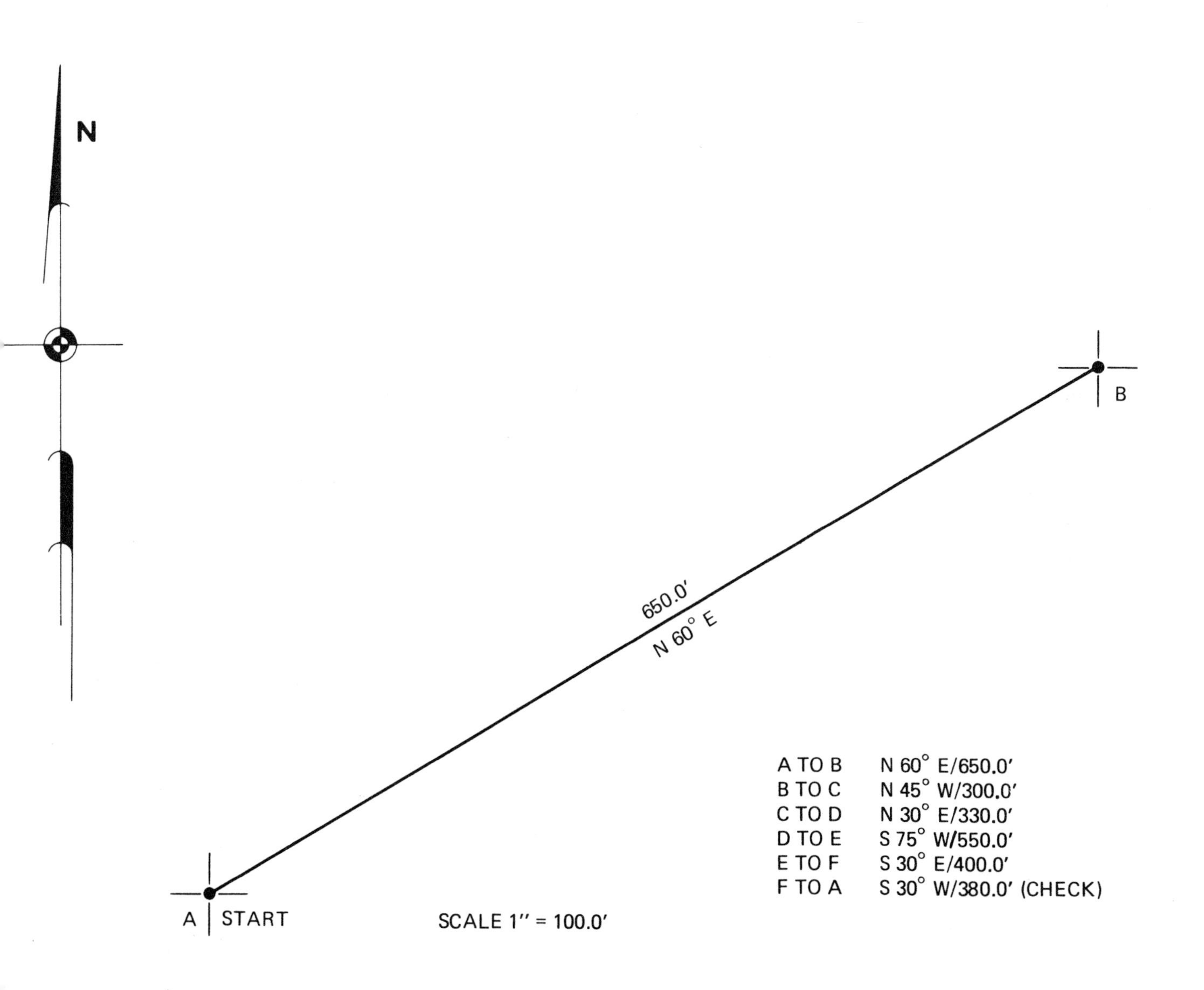

Practice Exercise 2-2

Using the land description given, carefully plot points A, B, C, D, E, F, and G. Draw a line from G to A to close the traverse. Compare your work to the answer on page 43. Keep this drawing as it will be used in exercise 2-9.

A TO B	820.0'/S 80° E
B TO C	740.0'/S 57° W
C TO D	1160.0'/S 45° E
D TO E	1030.0'/S 82° W
E TO F	475.0'/N
F TO G	525.0'/N 30° W
G TO A	/

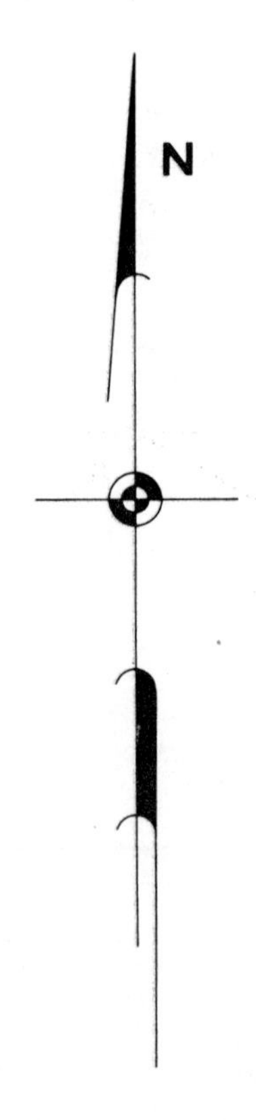

SCALE 1" = 200.0'

FINDING ANGLES BETWEEN BEARINGS

With One Quadrant

If there are two bearings such as those given in figure 2-8, the angle between the bearings can be calculated by adding the quadrant north to east. It is 80 degrees to the one bearing and 15 degrees to the other. Subtract 15 degrees from 80 degrees. The included angle between the bearings is 65 degrees.

Given:

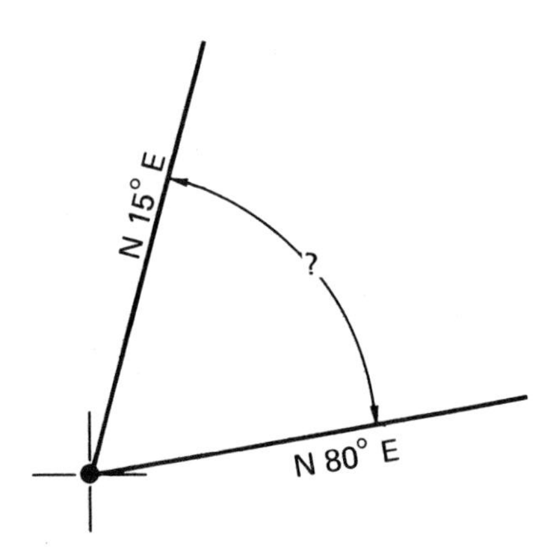

Method:

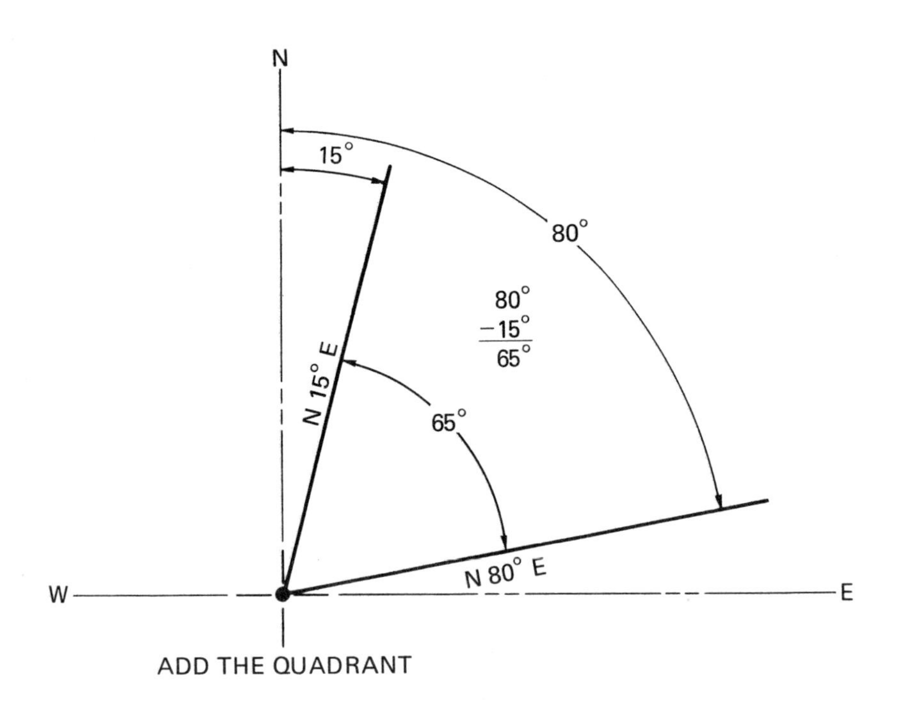

Fig. 2-8

Practice Exercise 2-3

Figure the angle between the bearings given. Compare your work to the answer on page 44.

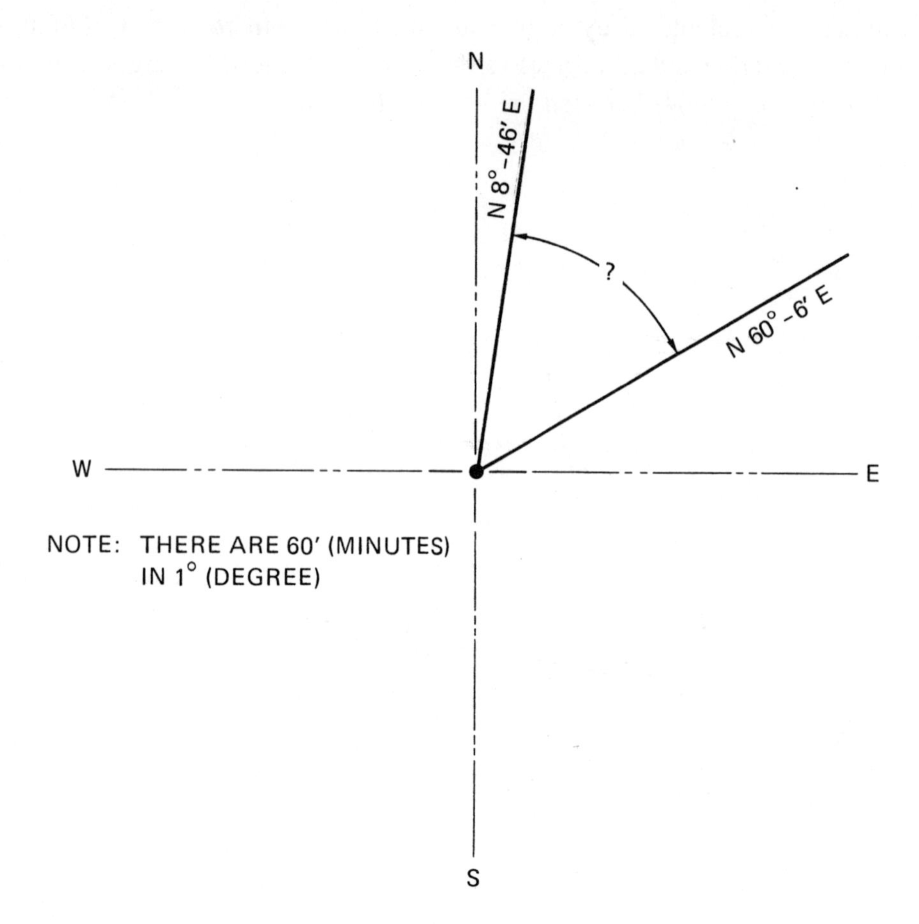

With Two Quadrants

If there are two bearings such as those given in figure 2-9, the two quadrants must be added together, north to south (180 degrees). Use the same method as in figure 2-8.

Given:

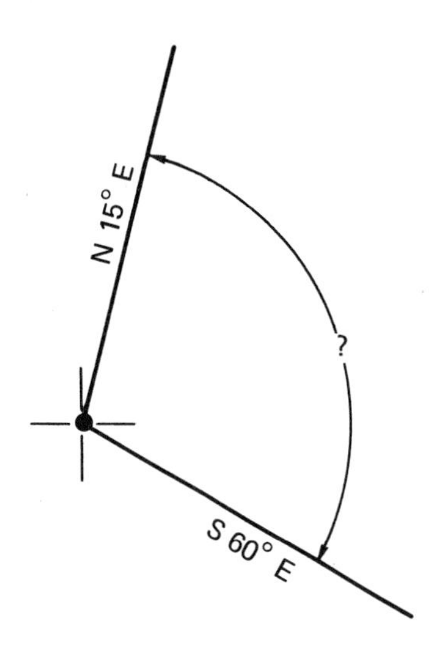

Method:

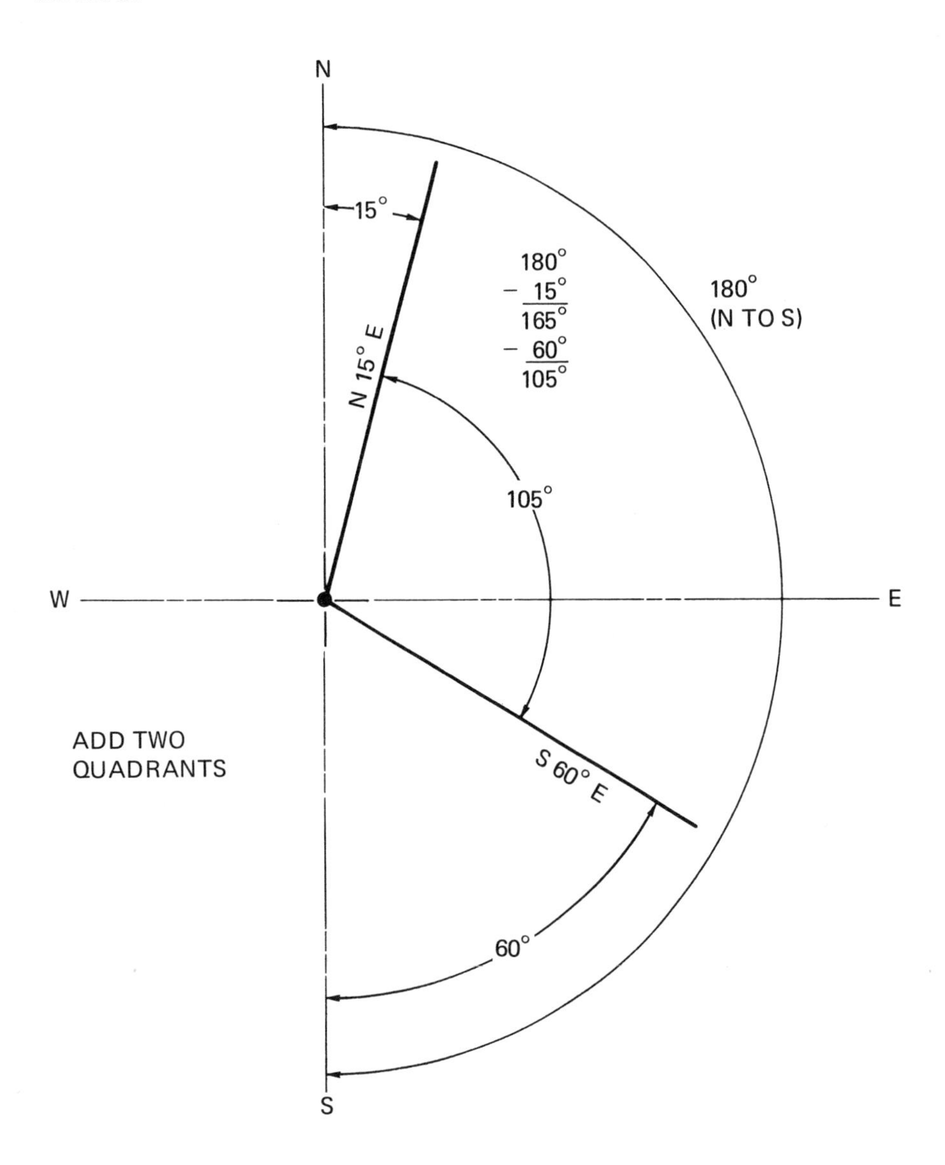

Fig. 2-9

Practice Exercise 2-4

Find the angle between the bearings given. Compare your work to the answer on page 44.

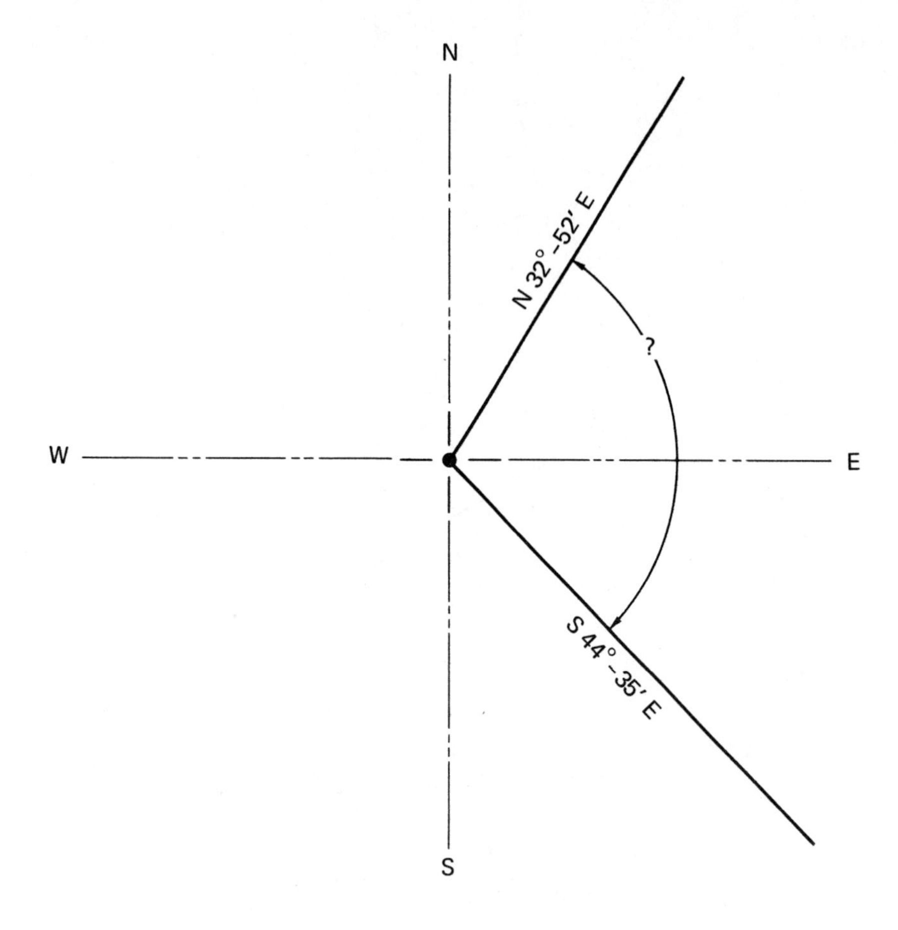

With Three Quadrants

If there are two bearings such as those given in figure 2-10, the three quadrants must be added. Add: 15° + 90° + 30° = 135°

Given:

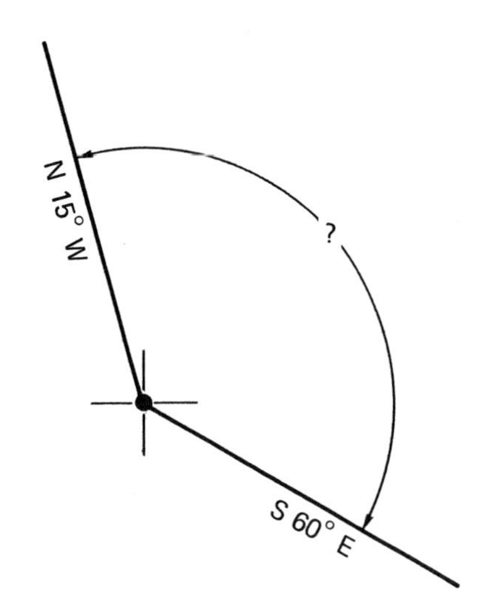

Method:

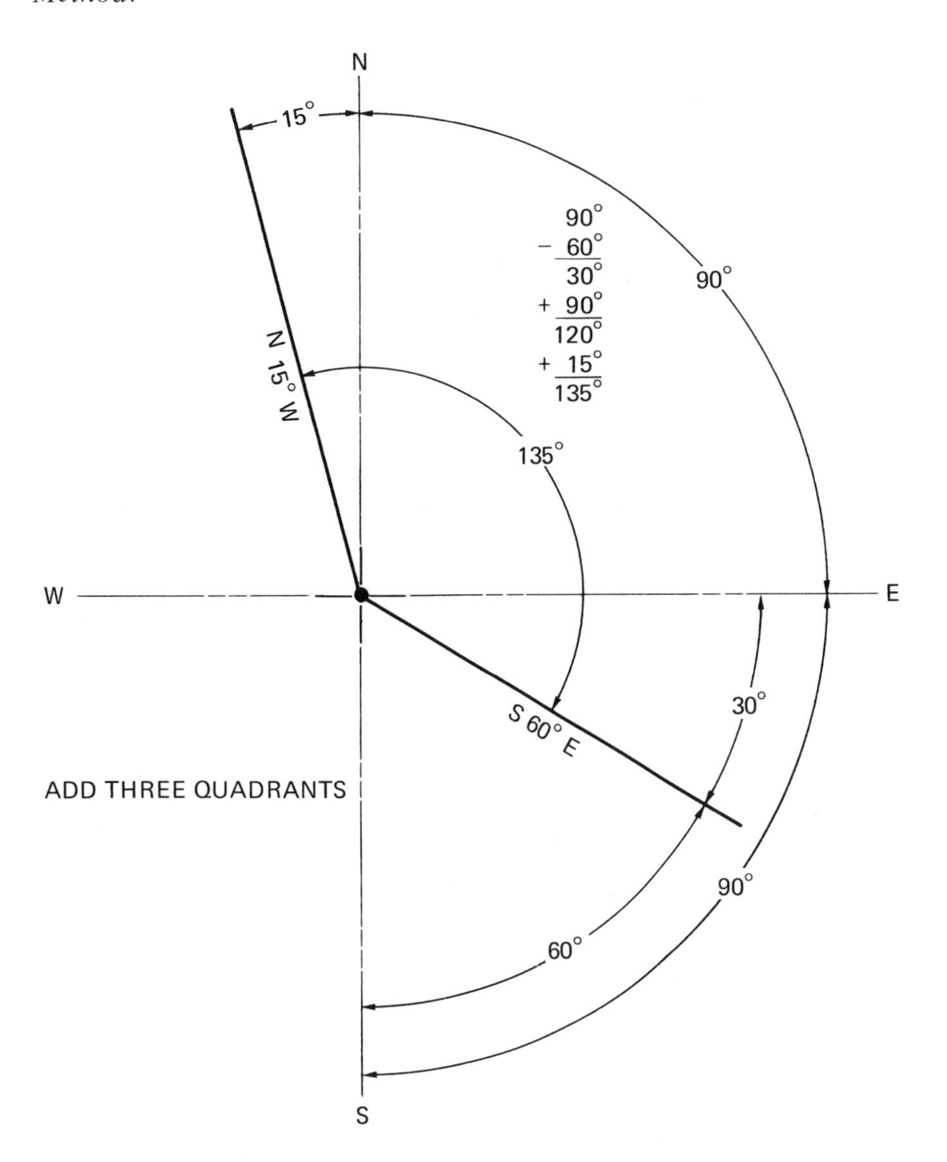

Fig. 2-10

Practice Exercise 2-5

Find the angle between the bearings given. Compare your work to the answer on page 44.

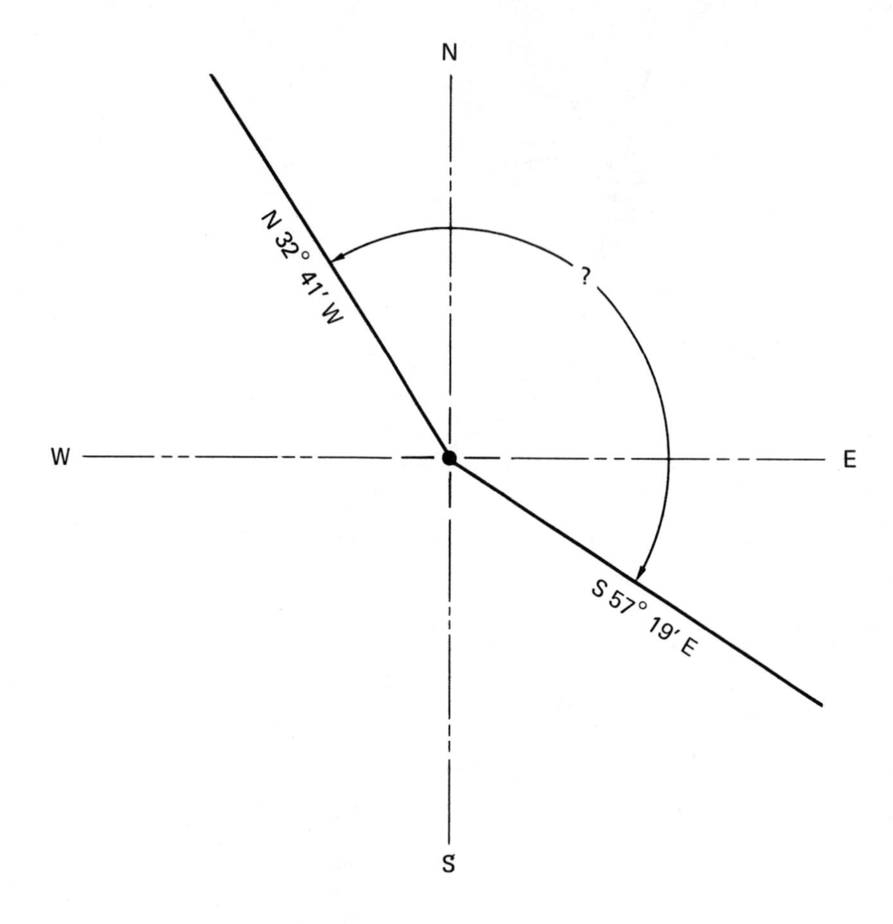

With Four Quadrants

If there are two bearings such as those given in figure 2-11, all four quadrants are added. Add: 15° + 180° + 30° = 225°

Given:

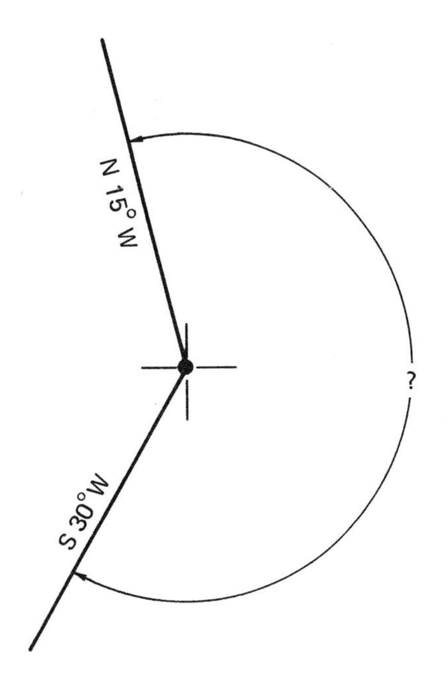

Method:

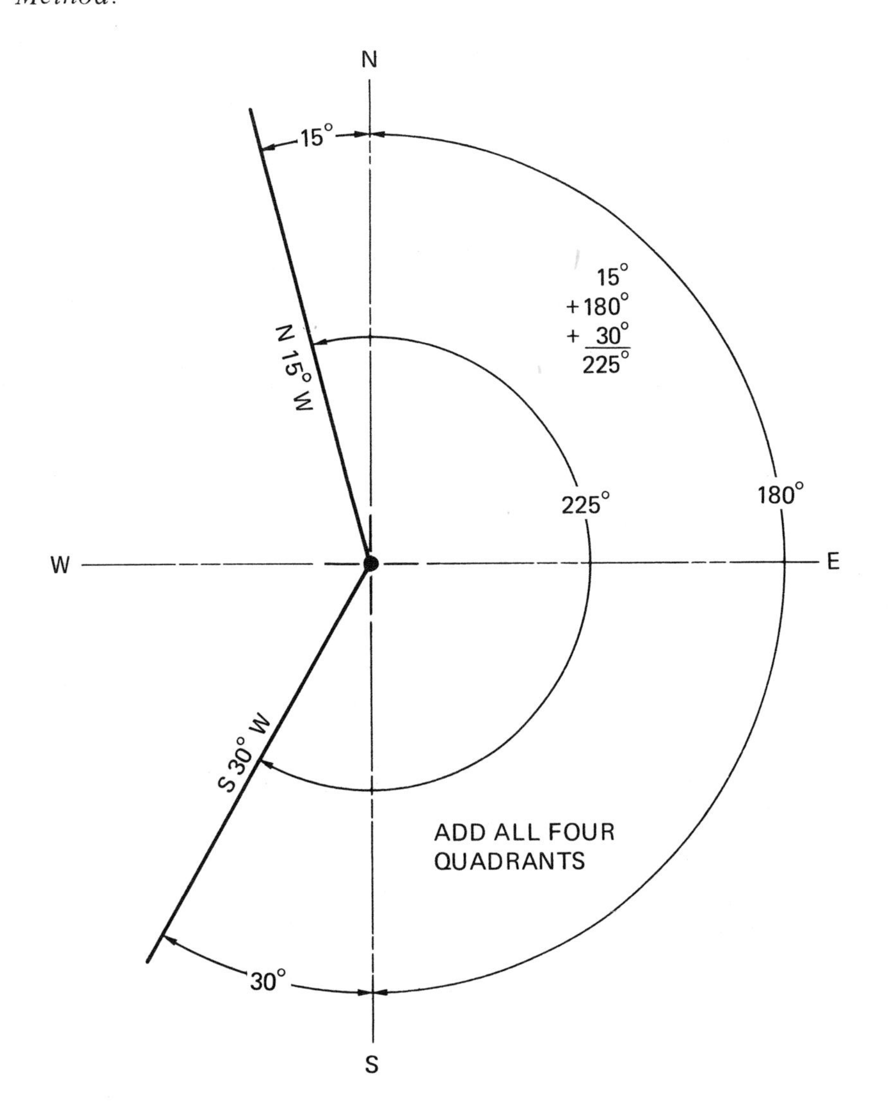

Fig. 2-11

Practice Exercise 2-6

Find the angle between the bearings given. Compare your work to the answer on page 44.

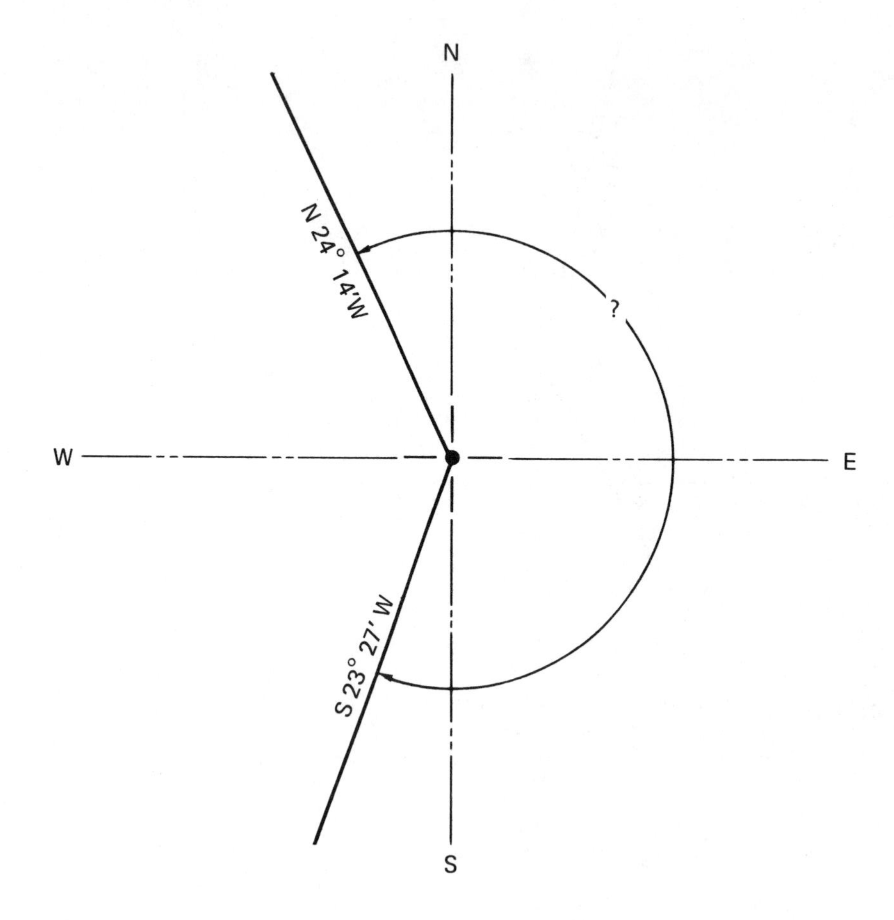

Practice Exercise 2-7

Calculate the angles in exercises A through F. Do all math beside each problem. Compare your work to the answers on page 44.

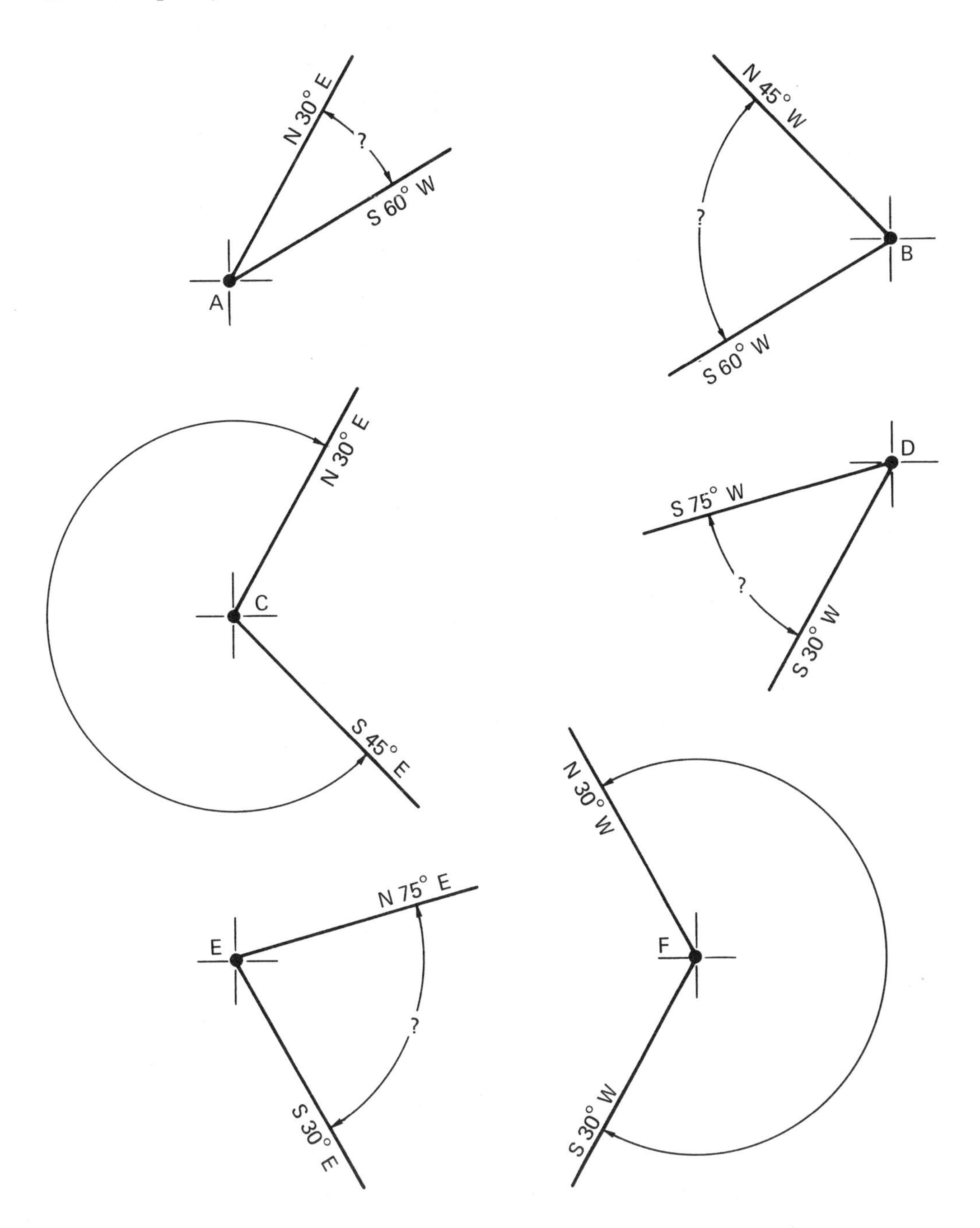

Practice Exercise 2-8

Add the inside angles to the drawing made in exercise 2-1. Include all distances and bearings. Compare your work to the answer on page 45.

Practice Exercise 2-9

Add inside angles to the drawing made in practice exercise 2-2. Make simple sketches of each corner so there is no chance of error. Compare your work to the answers on pages 46 and 47.

VERNIER

Study the vernier scale on a drafting machine. Inside there is a fixed scale graduated in full degrees; outside there is a vernier scale with twelve divisions of a degree, in five minute graduations, equaling one degree (12 x 5′ = 60′ = 1°).

Examples A through G show how to read a straight vernier scale graduated in inches and tenths of an inch. The vernier scale illustrated in figure 2-12 is read in exactly the same manner except that it is divided into degrees and minutes.

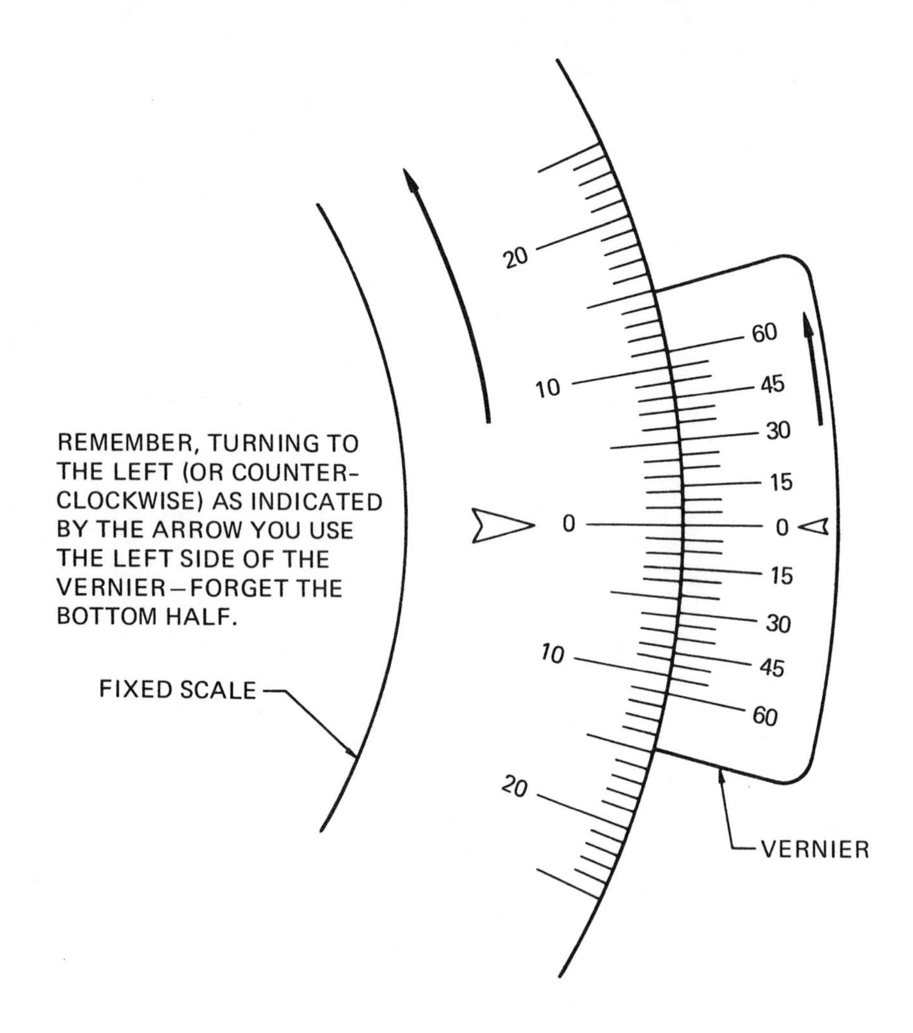

Fig. 2-12 Vernier scale

On the drafting machine's vernier scale, readings can be taken from either the right or left of 0 degree. If the angle turns to the right, only the right side of the scale is used.

The fractional part of any division can be read exactly by means of a vernier scale.

Example A: The arrow indicates the nearest *full* division, in this case "0."

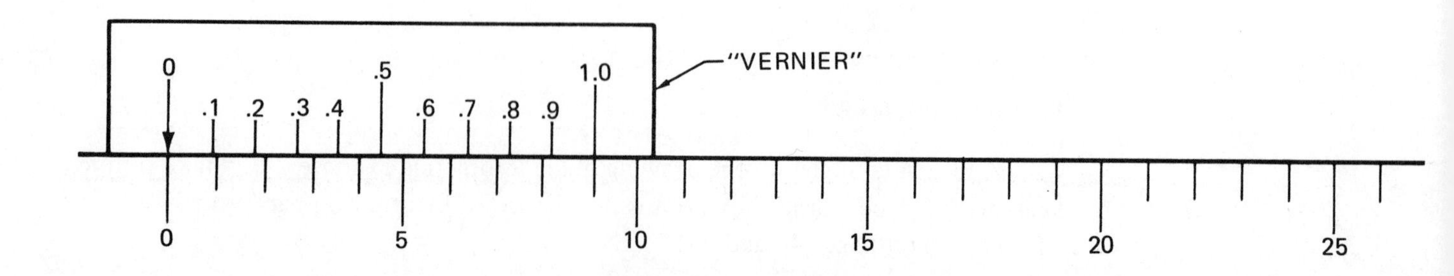

Example B: Moving from left to right, the arrow indicates 10, the nearest full division. On the vernier scale, find a line that lines up with another line; in this case it is .5. This reading is 10.5.

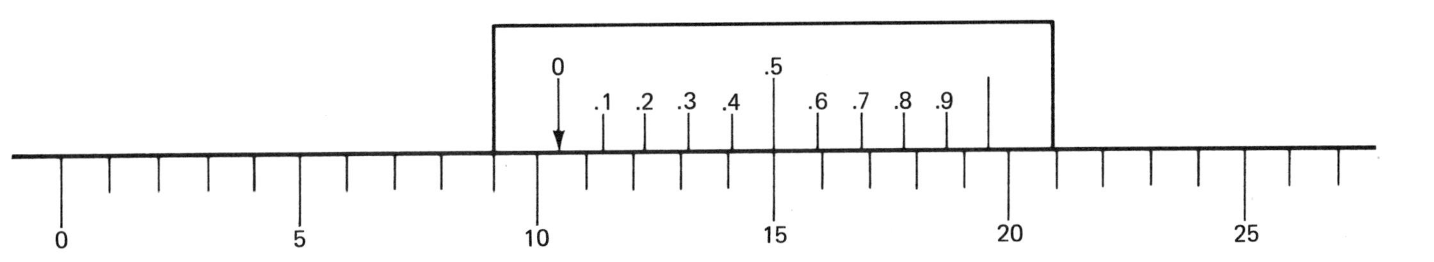

Example C: Moving from left to right, the arrow indicates 15. The .2 lines up with a line below. The reading is 15.2.

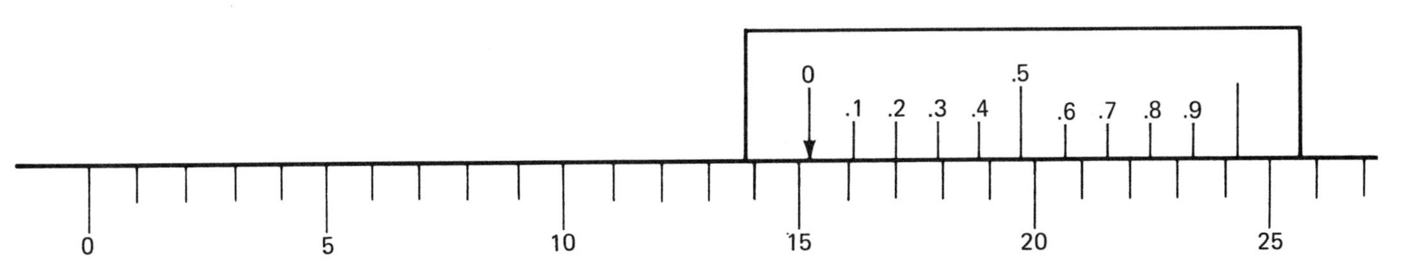

Example D: Moving from left to right, the arrow shows 7 full divisions with .4 lining up with another line. The reading is 7.4.

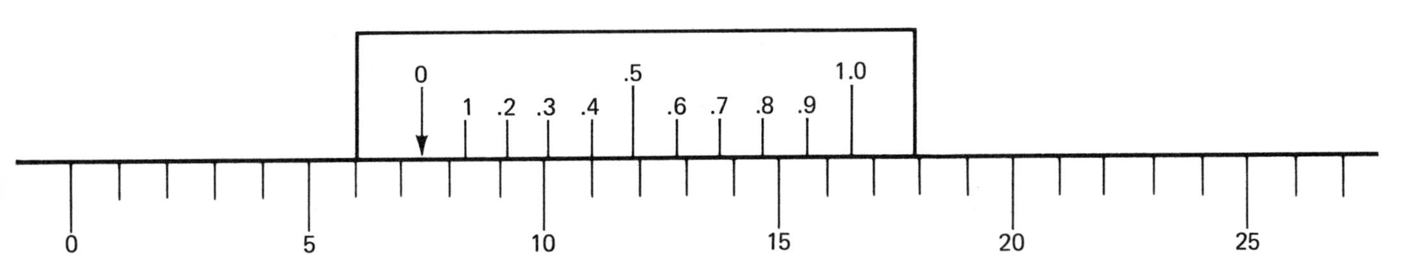

Example E: Fourteen full divisions with .6 lining up. Reading is 14.6.

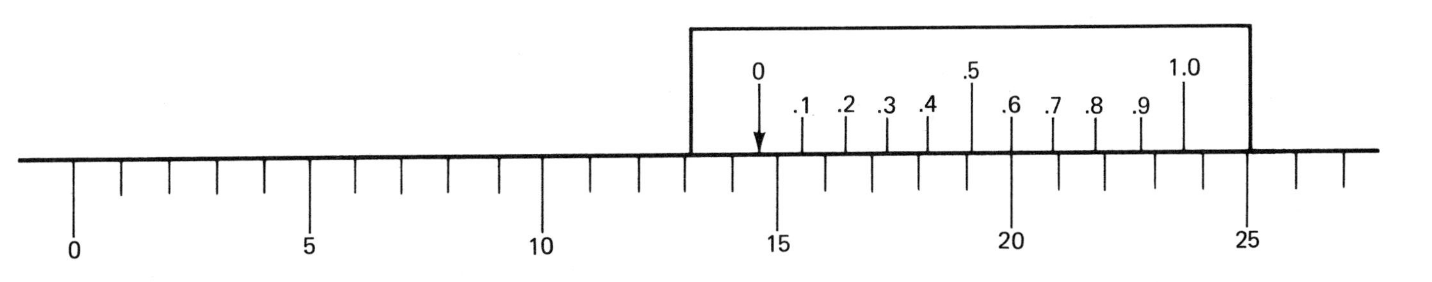

Example F: Moving from left to right, one full division; .8 fractional part. Reading = 1.8.

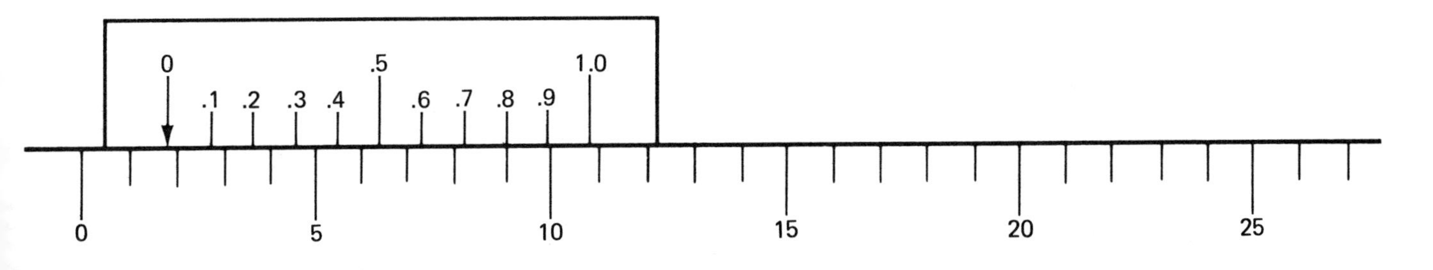

Example G: Eight full divisions; .1 fractional part. Reading = 8.1.

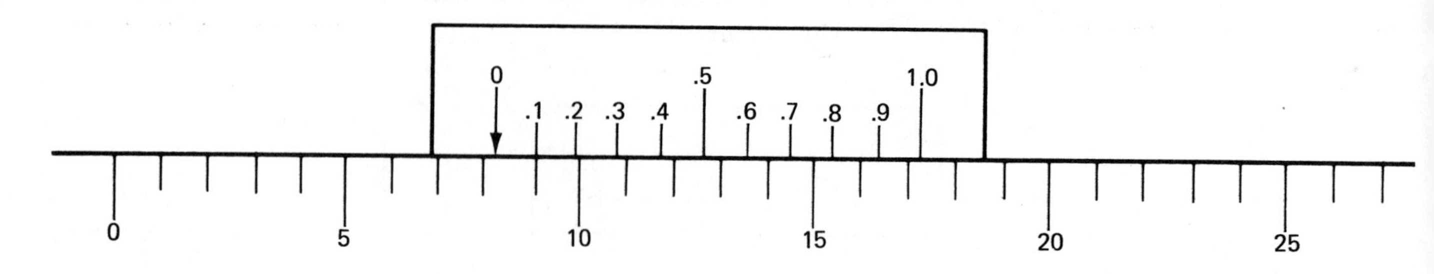

Practice Exercise 2-10

Complete each exercise. First locate the arrow and find the nearest full division. In the vernier scale, find a line that lines up with a line on the scale. Add it to the full division. Read each scale, record the answer, and compare your work to the answers on page 47.

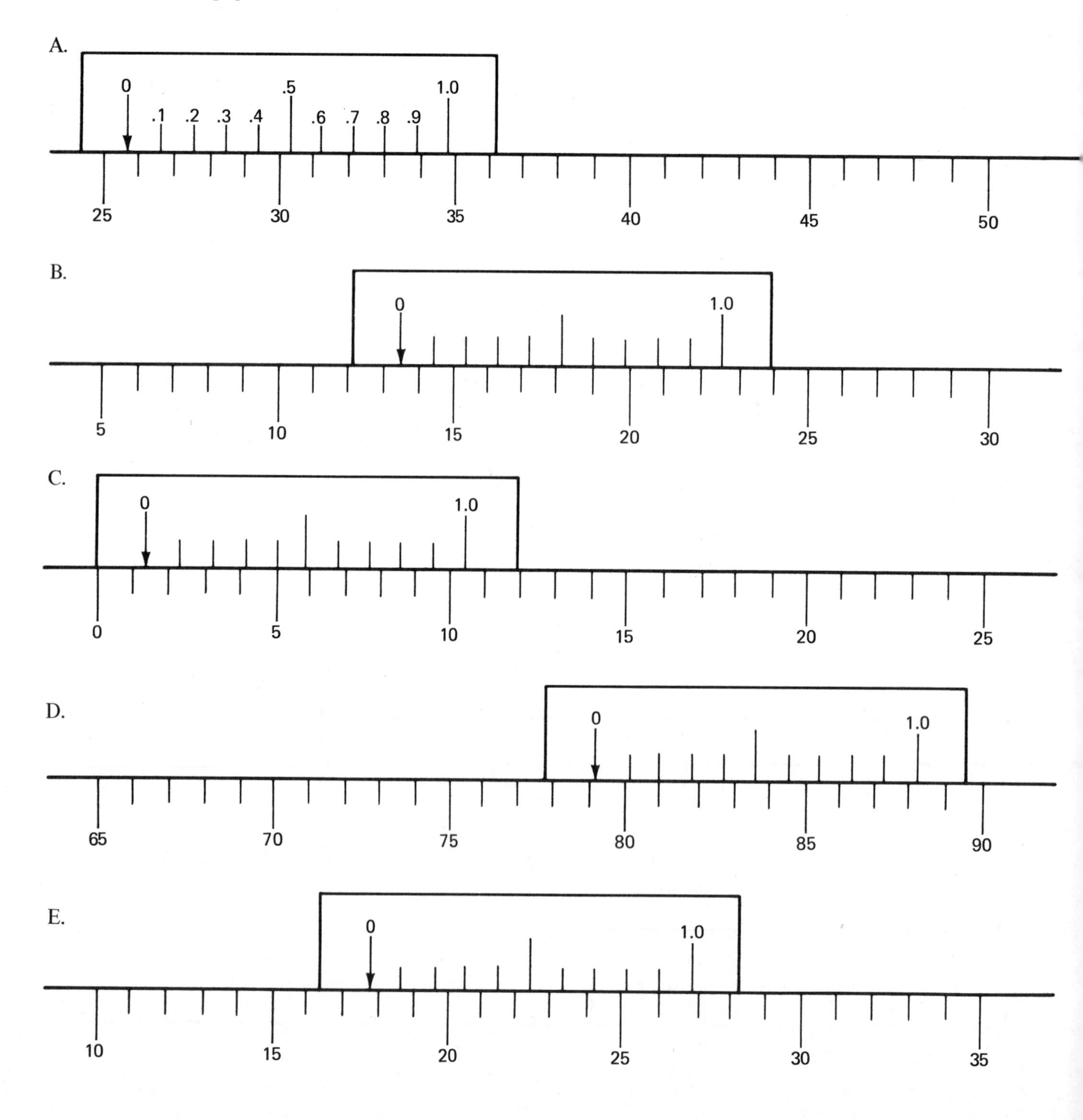

Practice Exercise 2-11

Starting from point A, plot the following in the space provided:

A–B N64°-30′E/520.0′
B–C S29°-15′E/400.0′
C–D S79°-45′W/450.0′
D–E S11°-55′W/390.0′
E-A ________/______

The answer must be within 10.0′ and/or 1°0′ of accuracy. Figure and add all inside angles. Scale: 1″ = 100.0′. Compare your work to the answer on page 48.

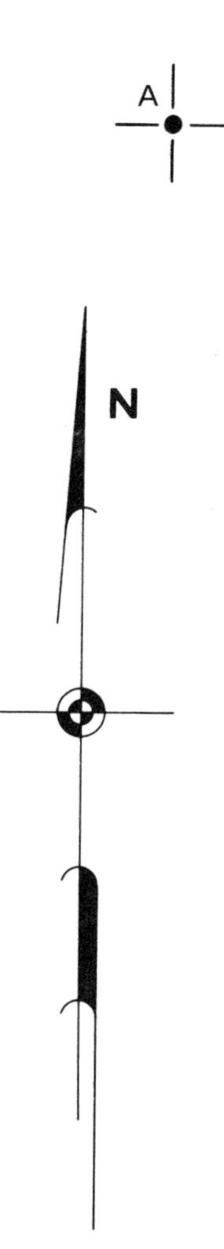

DEFLECTION PLOTTING

There are three ways to plot.

- Bearings
- Azimuths
- Deflection

Figure 2-13 illustrates *deflection*. Start with a point (point A) and have a heading (N 60° E) or some other point to head towards (point B). From then on it is simply measuring distances, finding points, and turning either left or right from the line of sight.

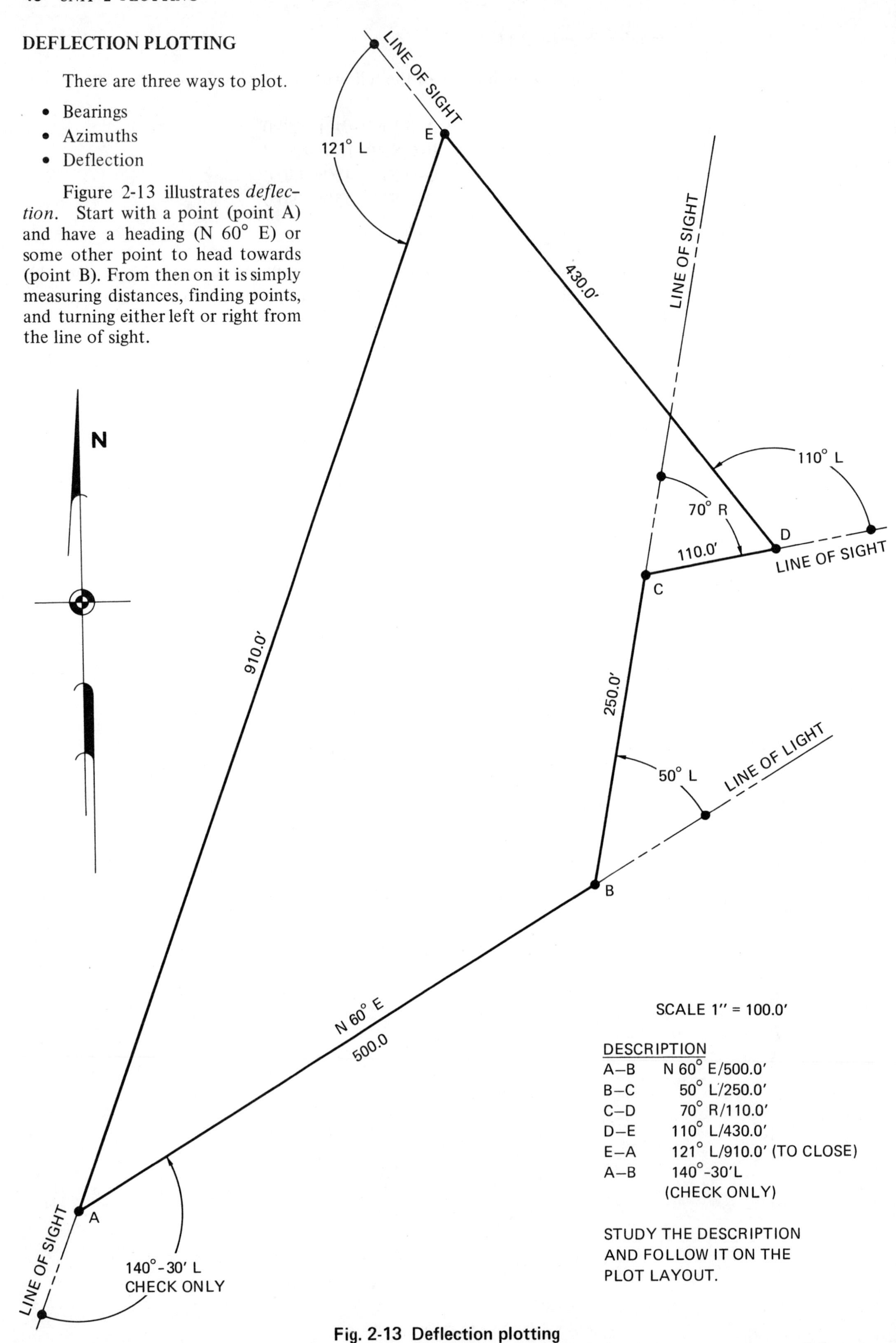

Fig. 2-13 Deflection plotting

Practice Exercise 2-12

Using the deflection angle method, plot from point A. Use either a protractor or a drafting machine to turn angles. Make a complete finished drawing of Lot #15. Compare your work to the answer on page 49.

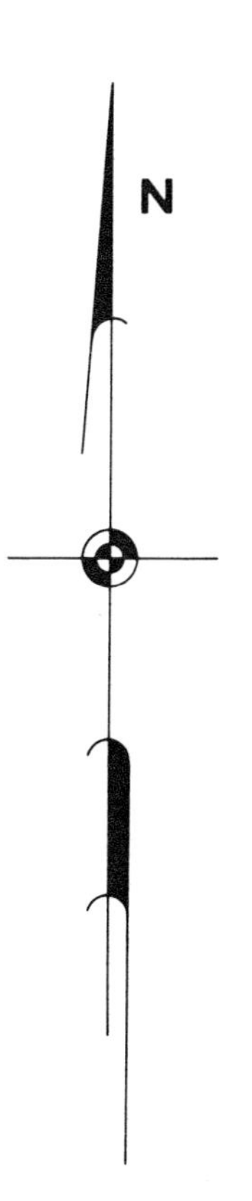

LOT 15

A ●

SCALE 1" = 100.0'

TYPICAL HOUSE-LOT LAYOUT

A–B	N 85° E/510.0'
B–C	86° L/380.0'
C–D	77° L/400.0'
D–A	______ / ______ (TO CLOSE)
A–B	CHECK ANGLE OF ______

ANSWERS TO PRACTICE EXERCISES

Carefully compare your work to the answer for each exercise. Refer any questions to the instructor.

Exercise 2-1

Note the corners, distances, and heading callouts. A drawing such as this should have line contrast in order to make it easier to read and understand. If your answer is off more than 10 feet on the distance and/or more than 10 degrees, do the drawing again.

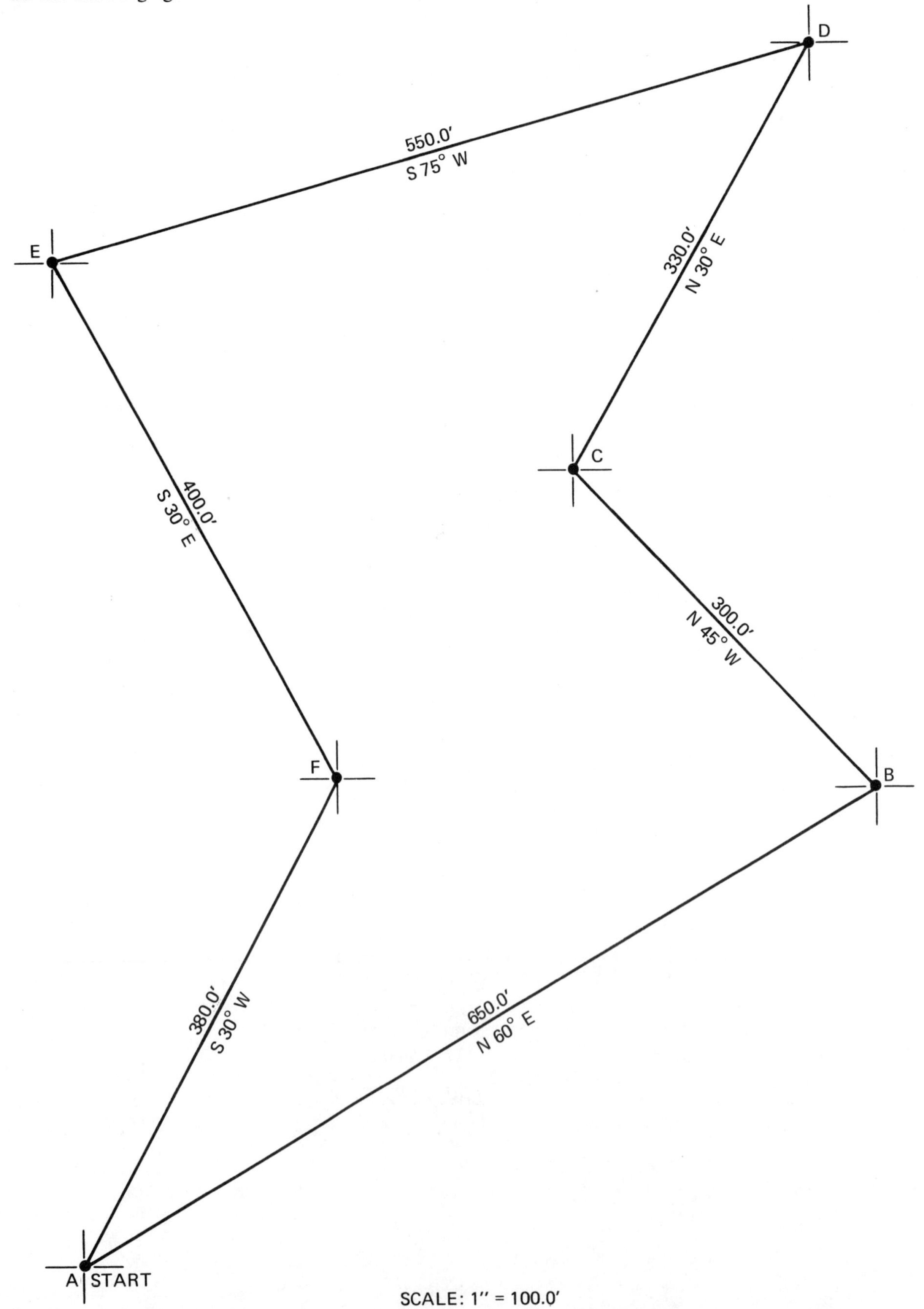

Exercise 2-2

Note each corner, distance, and heading callout. Check line weight. If the answer is off by more than 25 feet or 15 degrees, do the drawing again.

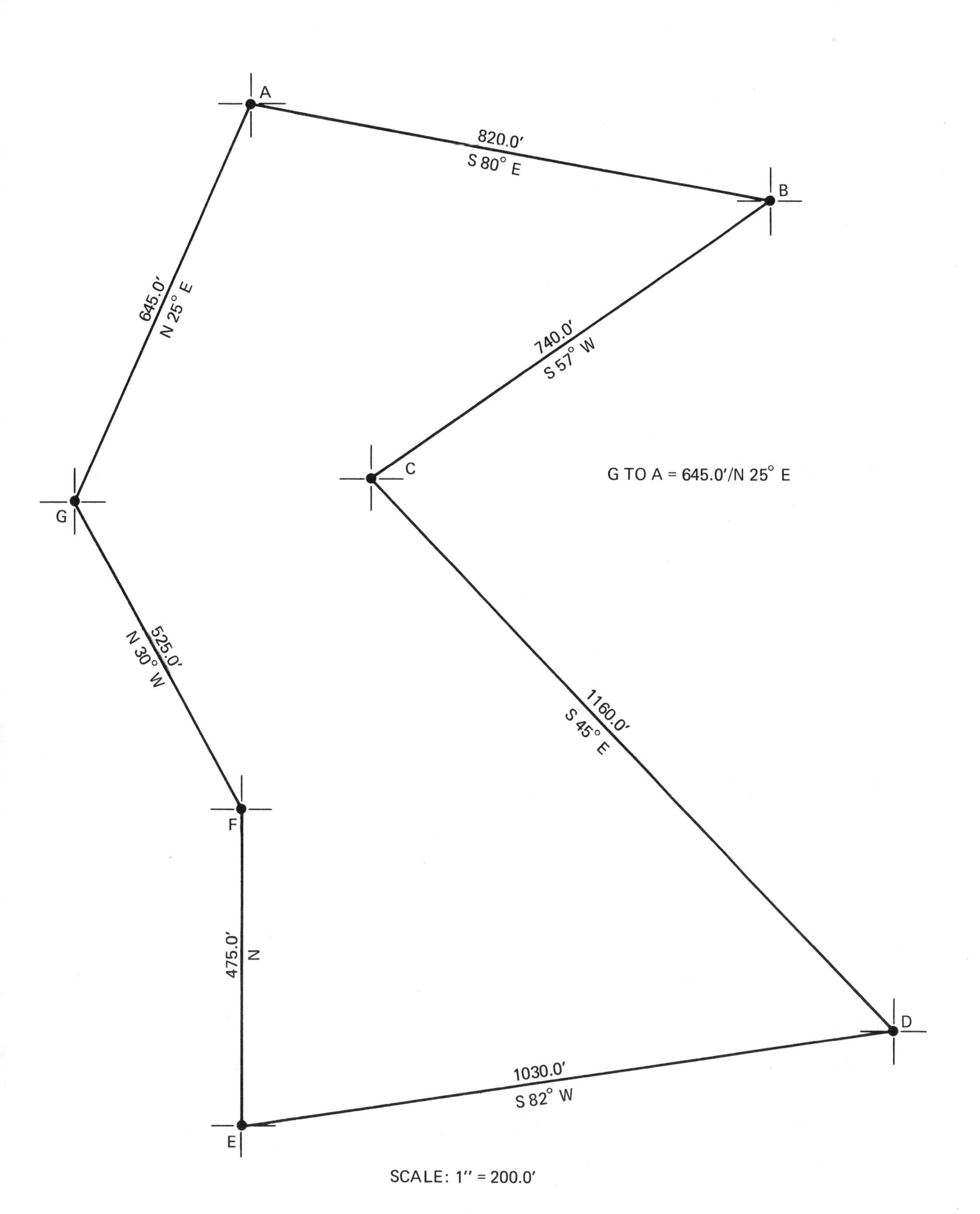

Exercise 2-3

$$\begin{array}{r} 60^\circ\text{-}06' \\ -\ \ 8^\circ\text{-}46' \\ \hline 51^\circ\text{-}20' \end{array}$$

Exercise 2-4

$$\begin{array}{r} 180^\circ\text{-}00' \\ -\ \ 32^\circ\text{-}52' \\ \hline 147^\circ\text{-}08' \\ -\ \ 44^\circ\text{-}35' \\ \hline 102^\circ\text{-}33' \end{array}$$

Exercise 2-5

$$\begin{array}{r} 90^\circ\text{-}00' \\ -\ \ 57^\circ\text{-}19' \\ \hline 32^\circ\text{-}41' \\ +\ \ 90^\circ\text{-}00' \\ \hline 122^\circ\text{-}41' \\ +\ \ 32^\circ\text{-}41' \\ \hline 155^\circ\text{-}22' \end{array}$$

Exercise 2-6

$$\begin{array}{r} 24^\circ\text{-}14' \\ +\ 180^\circ\text{-}00' \\ \hline 204^\circ\text{-}14' \\ +\ \ 23^\circ\text{-}27' \\ \hline 227^\circ\text{-}41' \end{array}$$

Exercise 2-7

A
$$\begin{array}{r} 60^\circ \\ -\ 30^\circ \\ \hline 30^\circ \end{array}$$

B
$$\begin{array}{r} 180^\circ \\ -\ \ 45^\circ \\ \hline 135^\circ \\ -\ \ 60^\circ \\ \hline 75^\circ \end{array}$$

C
$$\begin{array}{r} 30^\circ \\ +\ 180^\circ \\ +\ \ 45^\circ \\ \hline 255^\circ \end{array}$$

D
$$\begin{array}{r} 75^\circ \\ -\ 30^\circ \\ \hline 45^\circ \end{array}$$

E
$$\begin{array}{r} 180^\circ \\ -\ \ 75^\circ \\ \hline 105^\circ \\ -\ \ 30^\circ \\ \hline 75^\circ \end{array}$$

F
$$\begin{array}{r} 30^\circ \\ +\ 180^\circ \\ +\ \ 30^\circ \\ \hline 240^\circ \end{array}$$

Exercise 2-8

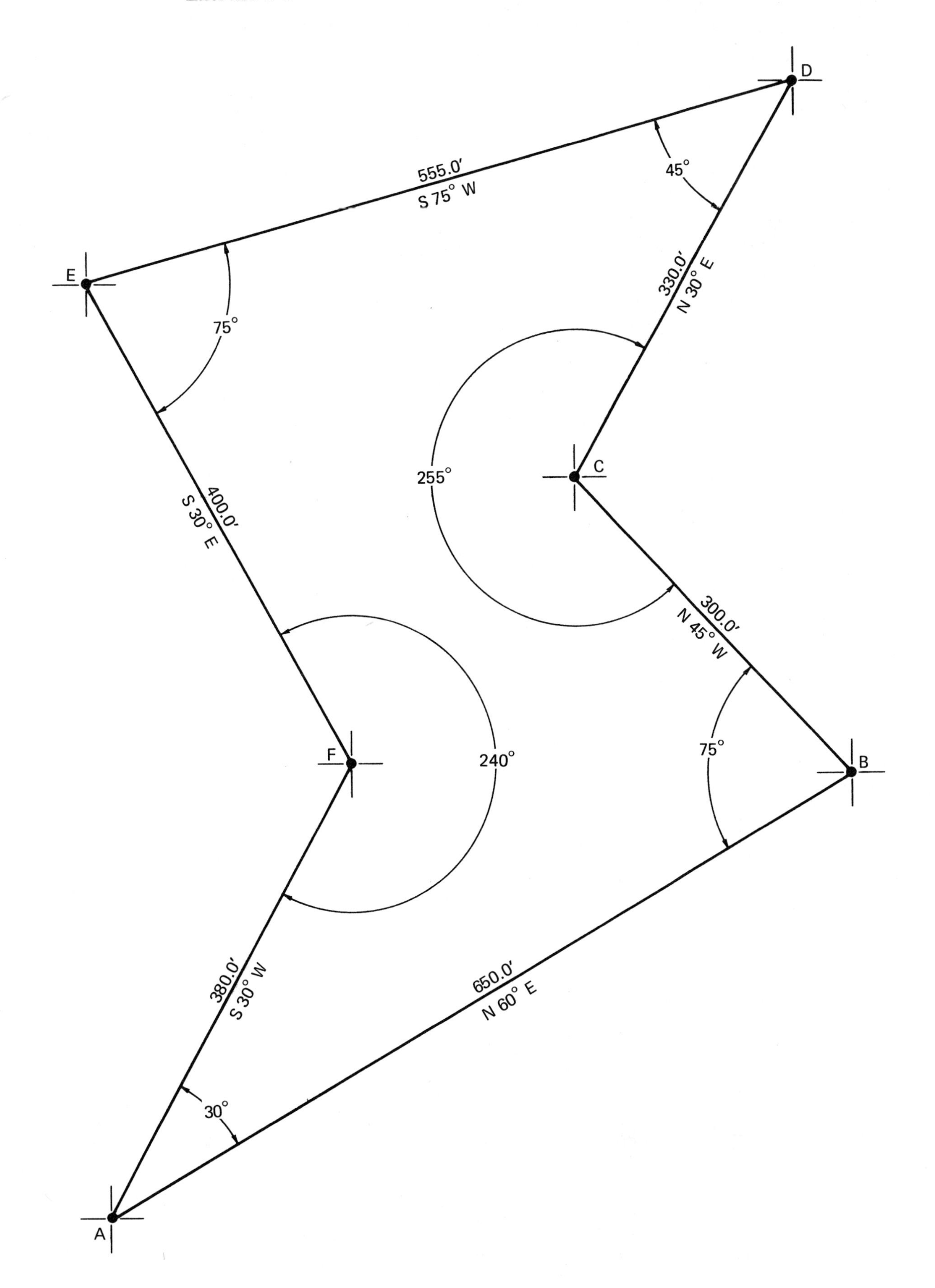
D
45°
555.0′
S 75° W
E
75°
330.0′
N 30° E
255°
C
400.0′
S 30° E
300.0′
N 45° W
F
240°
75°
B
380.0′
S 30° W
650.0′
N 60° E
30°
A

Exercise 2-9

Sketch neatly, rapidly, and approximately to scale. Sketching gives a visual idea of the given angle and helps eliminate errors.

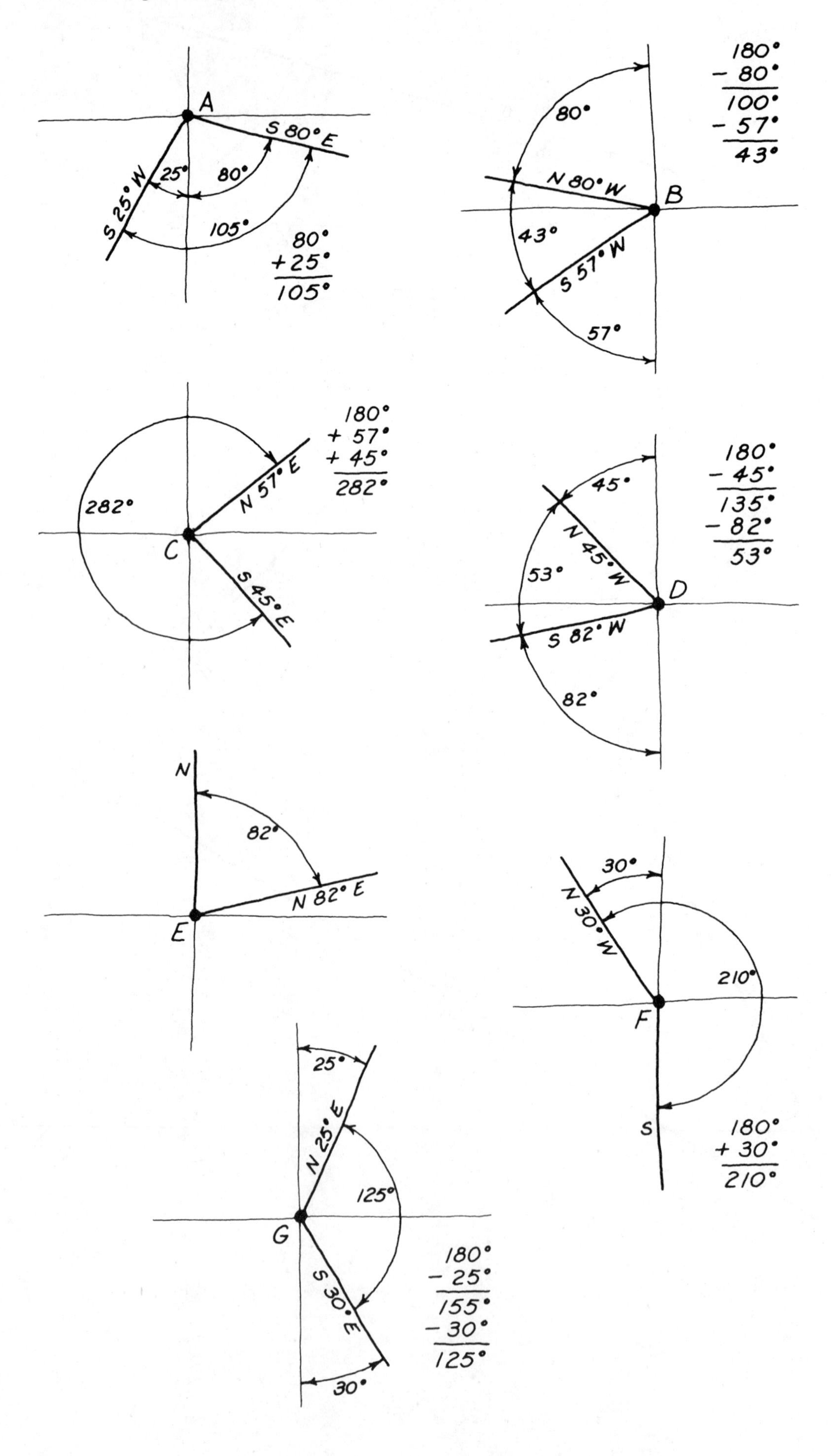

Add the inside angles that were calculated with the help of sketches to the drawing.

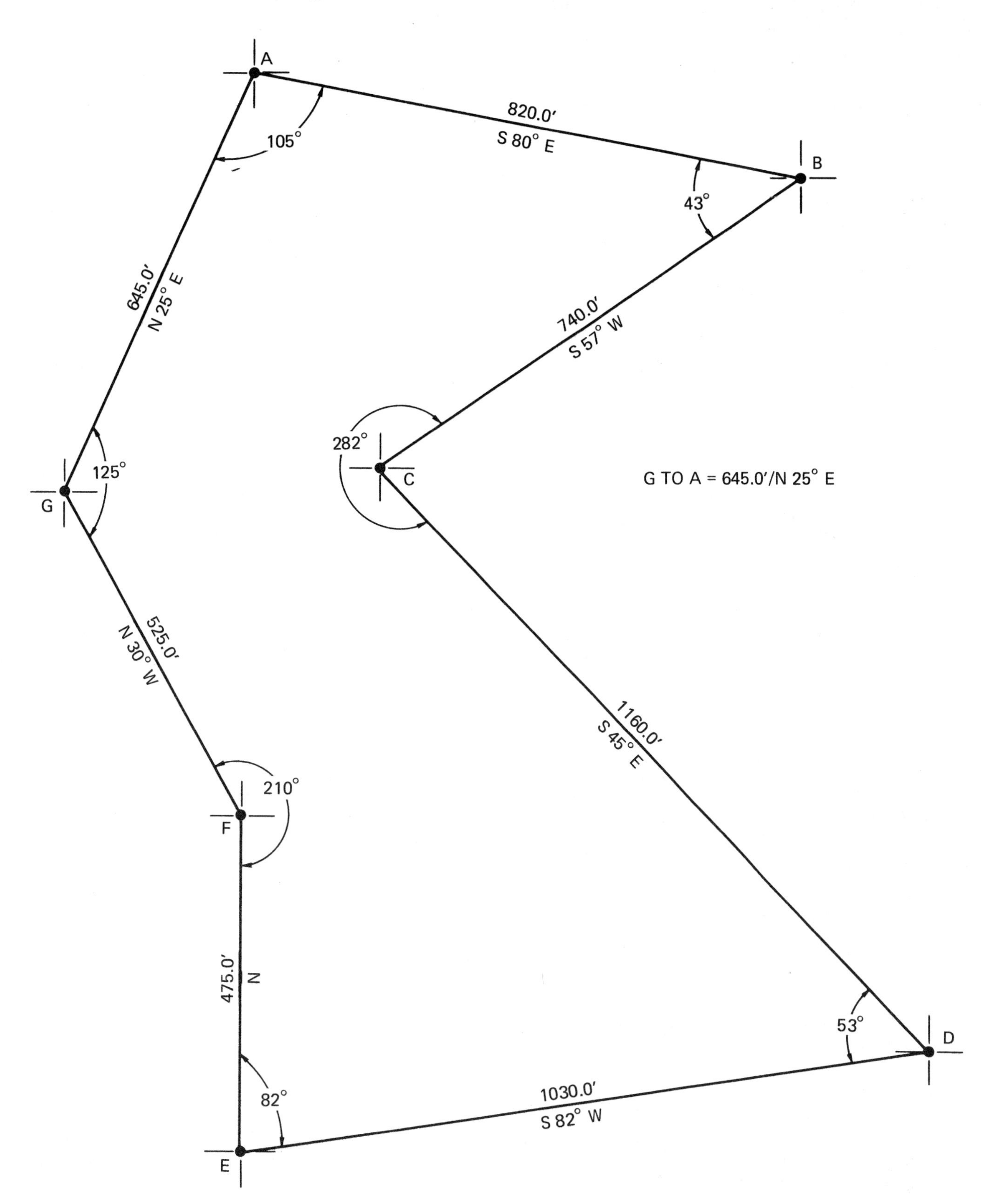

Exercise 2-10

A. 25.8
B. 13.6
C. 1.4
D. 79.2
E. 17.9

Exercise 2-11

The drawing should look exactly like this. It should be close to 10.0′ and within 1°-0′ of the answer E to A. Note the various line weights. Check each angle and distance heading.

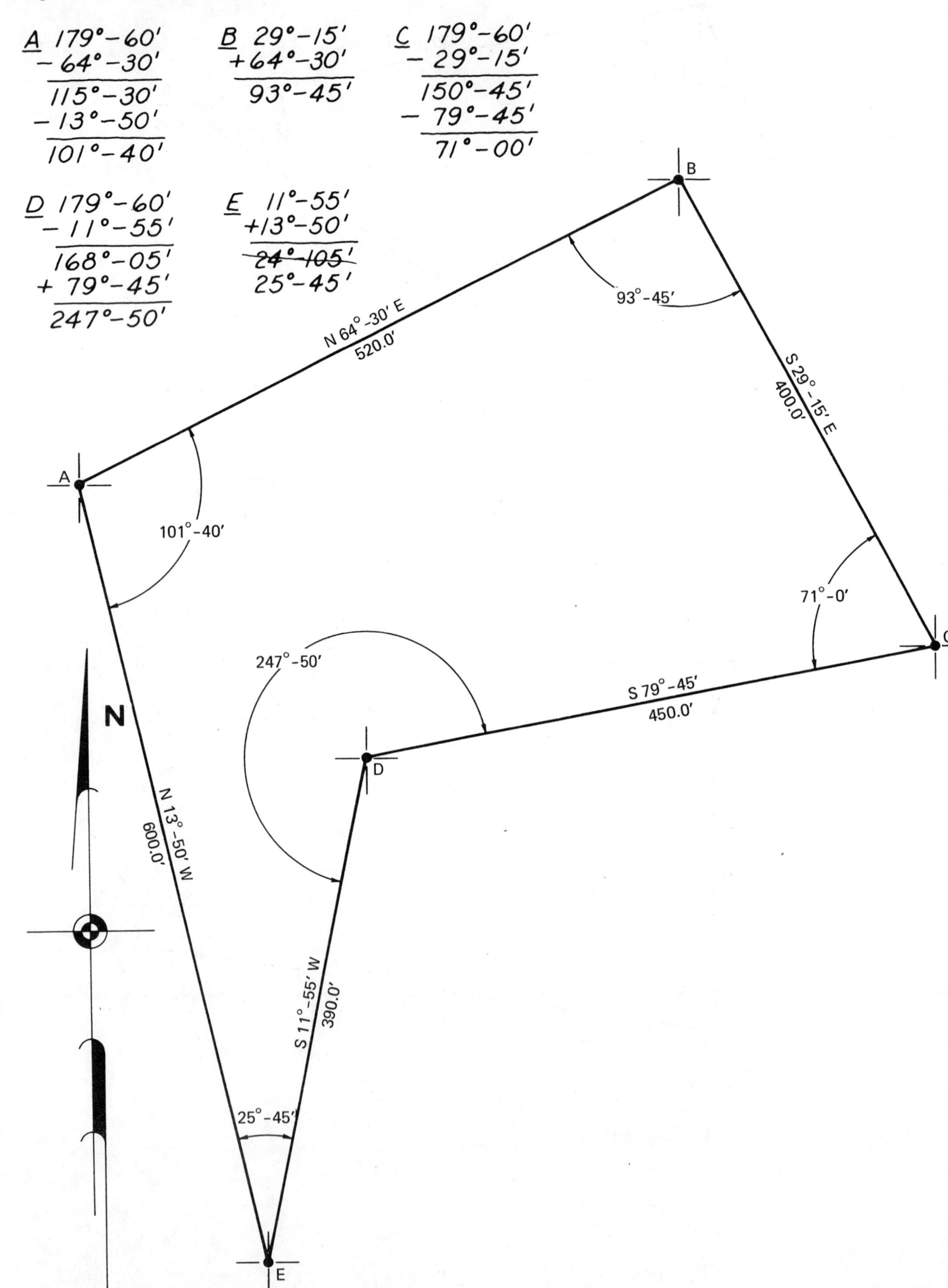

Exercise 2-12

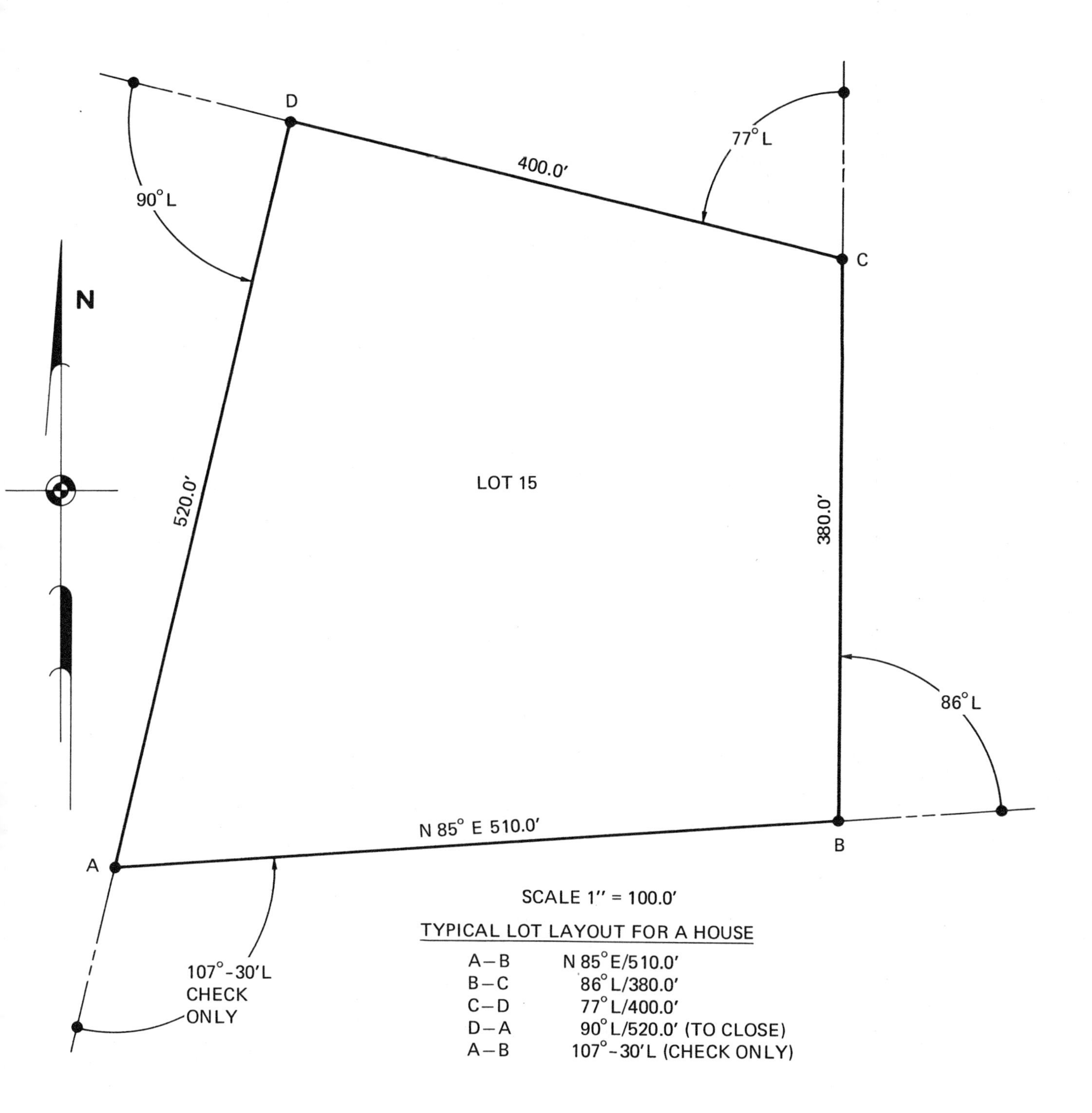

UNIT REVIEW

90-minute time limit

Make a complete finished drawing of the headings and distances given. Add all inside angles. Use correct line weight.

Scale: 1″ = 300.0′

Headings/Distances

A-B = N39°–00′E/1050.0′
B-C = S18°–30′E/880.0′
C-D = N65°–00′E/1100.0′
D-E = S 9°–45′W/2360.0′
E-F = N75°–00′W/1620.0′
F-G = N28°–15′E/1050.0′
G-A = ________ / ________

Before proceeding to the next unit:

______ Instructor's approval

______ Progress plotted

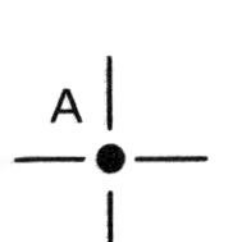
A

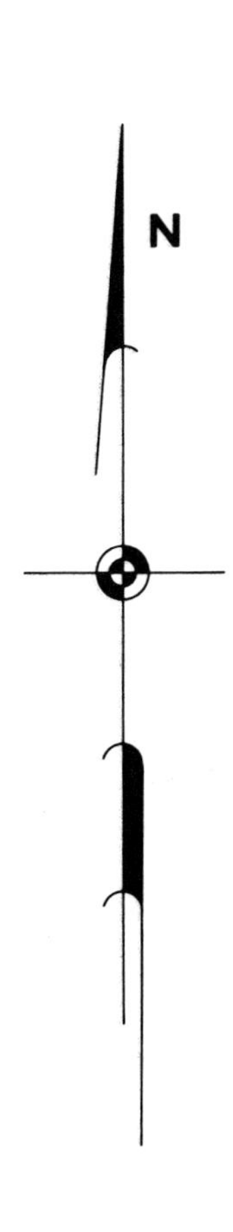
N

UNIT 3

SURVEYING

OBJECTIVE

The student will learn how to gather surveying information.

PRETEST

No time limit

Problem 1

Determine the following rod readings to the nearest hundredth of a foot. Record the answers in the spaces provided.

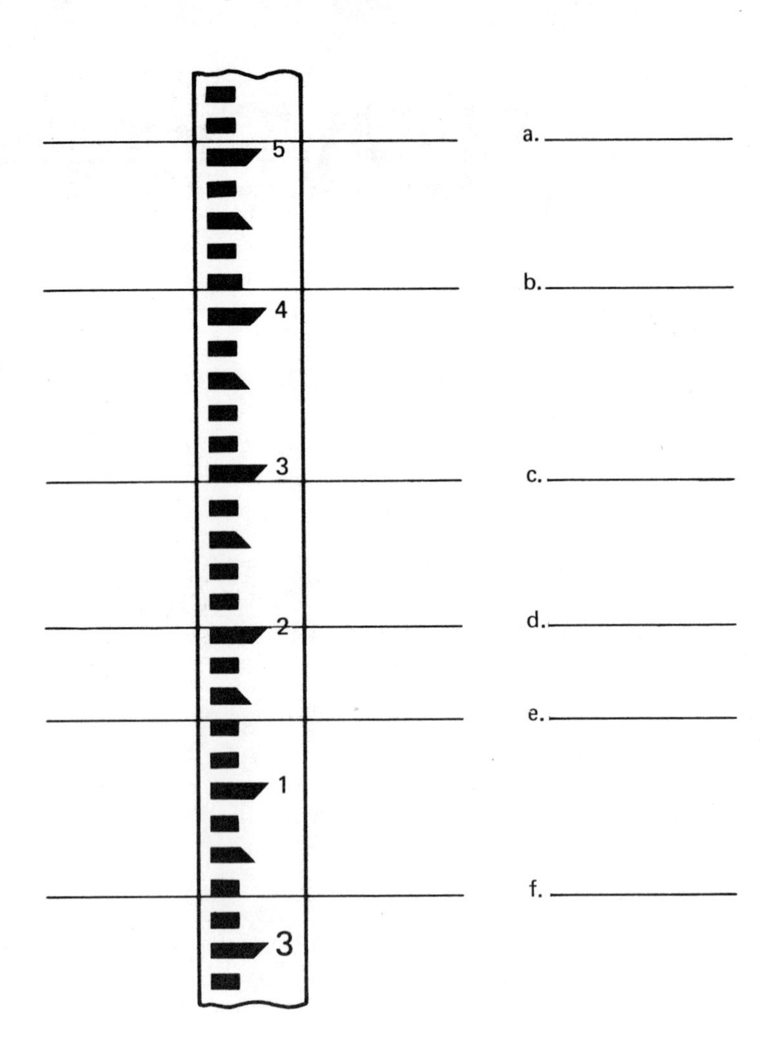

Problem 2

Using a stereoscope, locate the highest point on the photograph. Place an "X" on the high point.

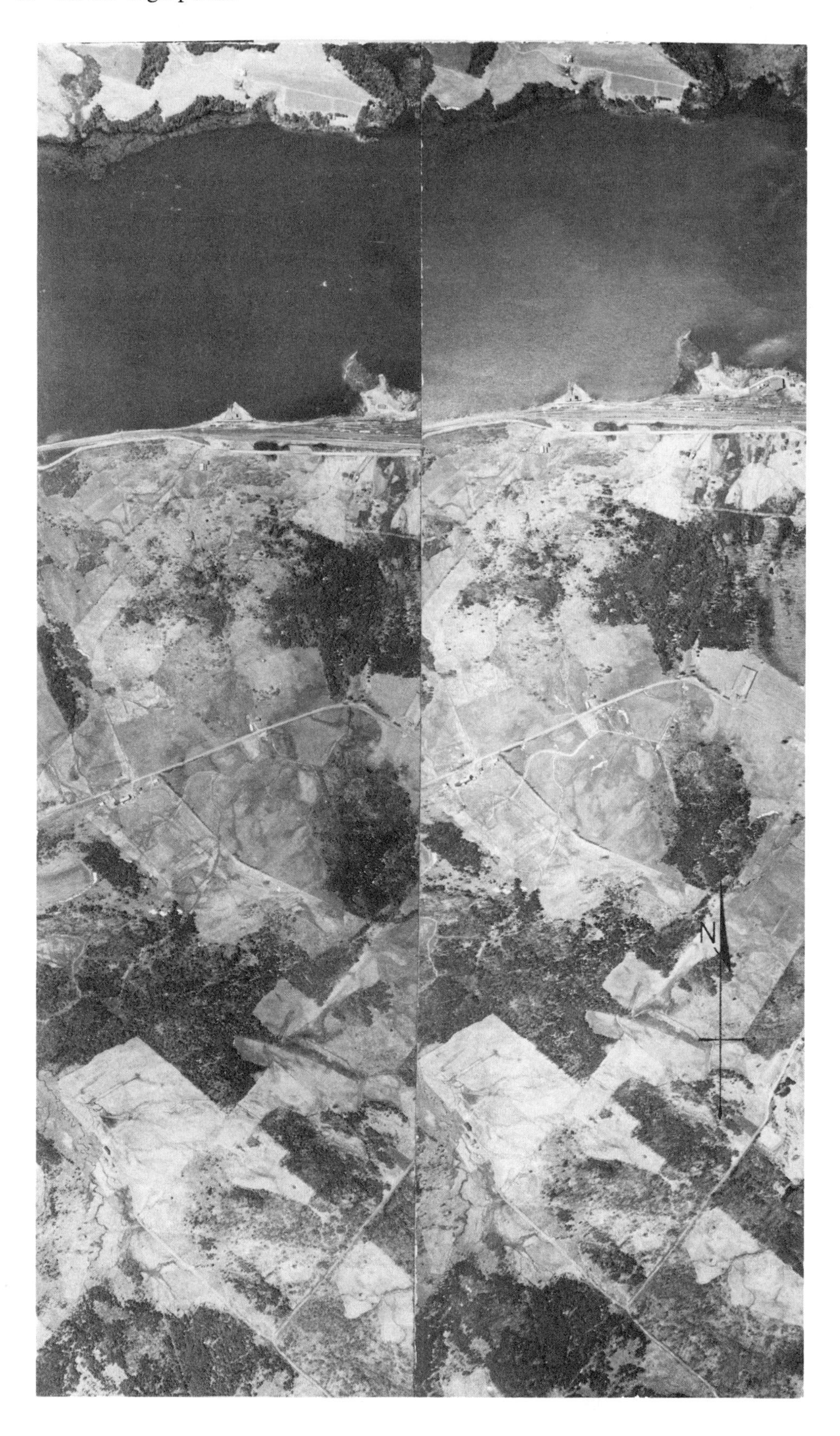

RELATED TERMS

Give a brief definition of each term as progress is made through the unit.

Party chief ____________________

Rear chainman ____________________

Head chainman ____________________

Survey crew ____________________

Breaking the chain ____________________

Stadia ____________________

Leveling ____________________

Bench mark ____________________

Turning point ____________________

Aerial photo ____________________

Photomosaic ____________________

Stereoscope ____________________

MEASURING IN THE FIELD

A survey crew is a group of people who work together to take the necessary measurements for a survey map. The leader of the crew is called the party chief. Working for the party chief are a clearing crew, a head chainman, a rear chainman, an instrument person, and a recorder or record keeper. Sometimes workers in a survey crew perform more than one duty.

The head chainman proceeds along the line to be measured carrying the head of the steel tape, called a *chain,* figure 3-1. The rear chainman follows with the end of the chain which has the higher numbers. When the tape is in position for measuring the line, the rear chainman calls out the even foot distance. The head chainman calls out the fractional part of a foot. The recorder marks the measurement in a log.

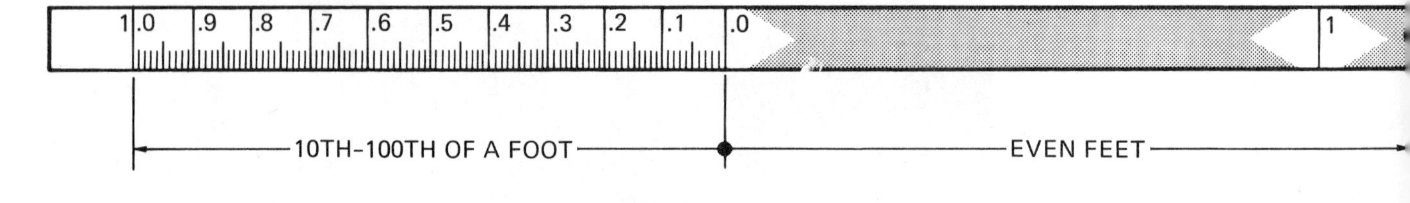

Fig. 3-1 Chain

The following callouts are made by the surveying crew for the measurement shown in figure 3-2.

Rear: "26" (indicates even footage at the rear of the tape)
Head: "Point 87" (indicates part of a foot at the front of the tape)
Rear: "26 point 87" (indicates total footage)
Head: "Check" (indicates agreement)

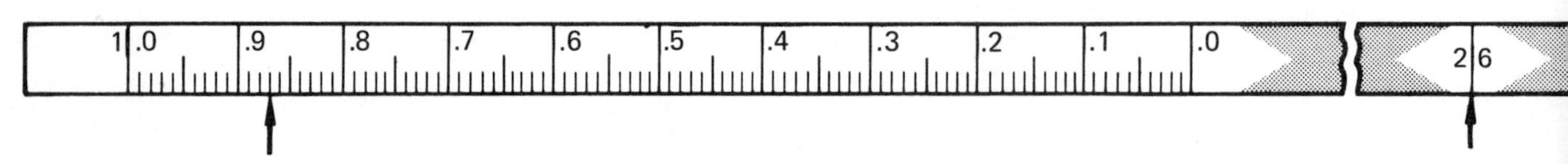

Fig. 3-2 Chain measurement of 26.87 feet

Figure 3-3 illustrates this conversation between the workers at the head and the rear of the chain.

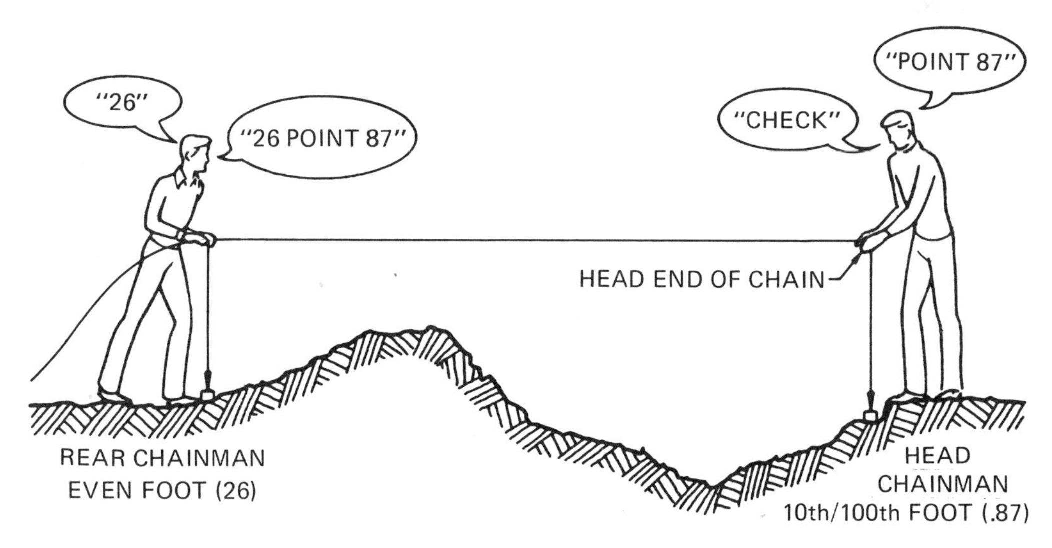

Fig. 3-3

In chaining, the workers must:

1. Measure in a straight line.

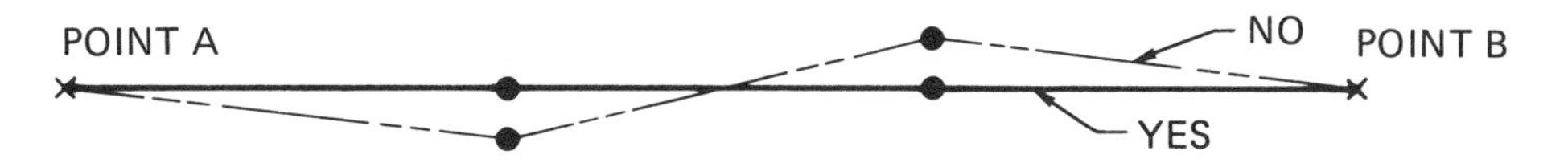

2. Pull at a standardized tension. The chain must be supported to prevent sag.

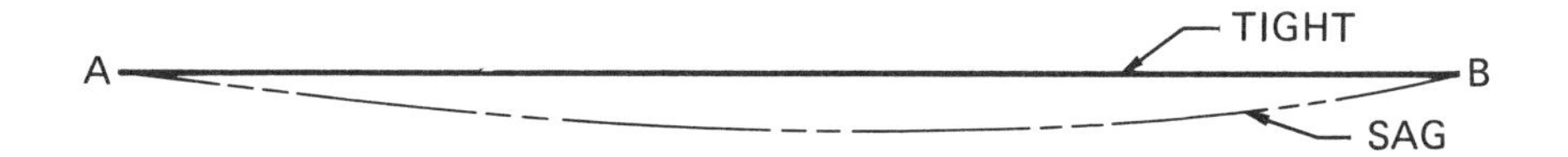

3. Hold the chain horizontally.

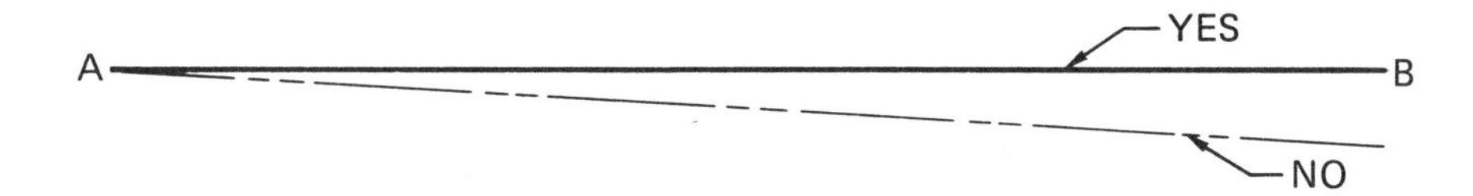

4. Allow for temperature.

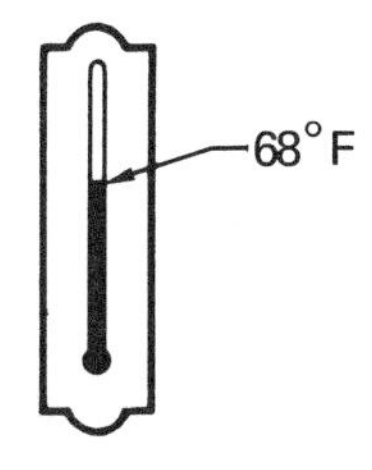

5. Call off measurement correctly, see figure 3-3.

Errors in taping may be classified by the following:

1. Incorrect length of tape
2. Pull or tension not consistent
3. Sag due to weight and wind
4. Poor alignment
5. Tape not horizontal
6. Temperature other than the standard 68° F
7. Improper plumbing
8. Faulty marking
9. Incorrect reading or interpolation

Breaking the Chain

The chain must be held horizontal even while chaining up or down a hill. Usually the chain is held between waist and chest level. This is accomplished while measuring on a slope by taking only short distance measurements, figure 3-4. This procedure is called *breaking the chain.*

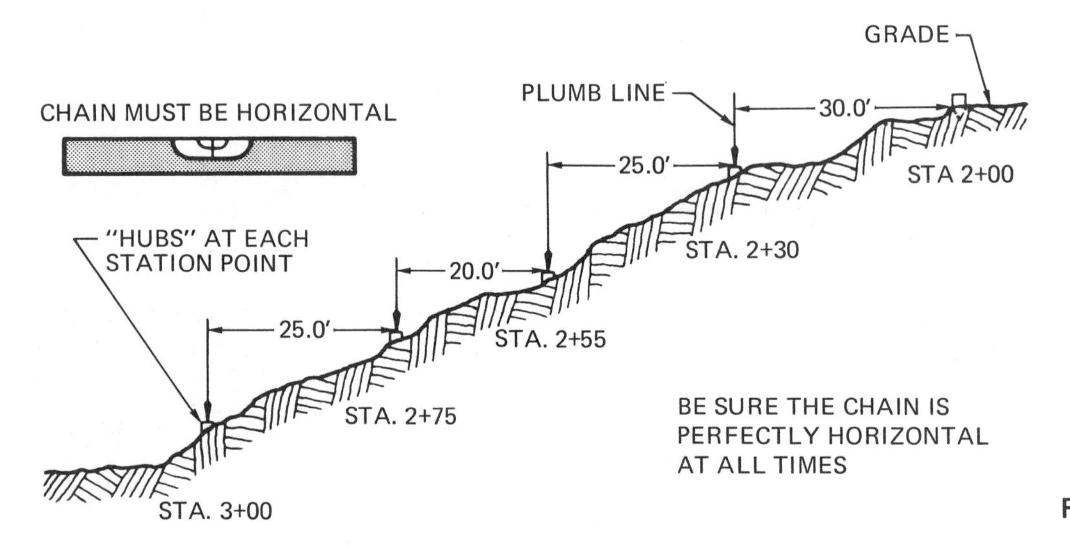

Fig. 3-4 Breaking the chain

Stadia Theory

The term *stadia* is a Greek word for a unit of length. It was originally a distance in an athletic contest that equalled approximately 600 feet. Stadia now refers to the top and bottom cross hairs that are aligned with an object through a transit sight, figure 3-5. A *transit* is a basic surveying instrument used to measure horizontal and vertical angles.

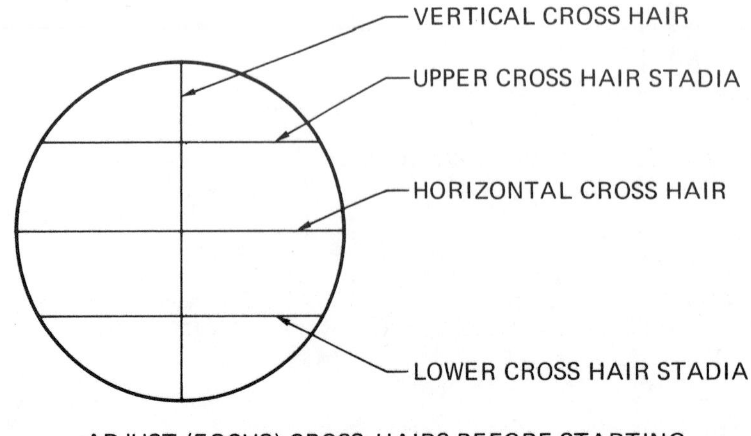

Fig. 3-5 Cross hairs in a transit sight

The stadia theory is illustrated in figure 3-6. Set the level, though it need not be exact. Then adjust the bottom cross hair on an even foot mark and lock. Read the top cross hair to calculate distance.

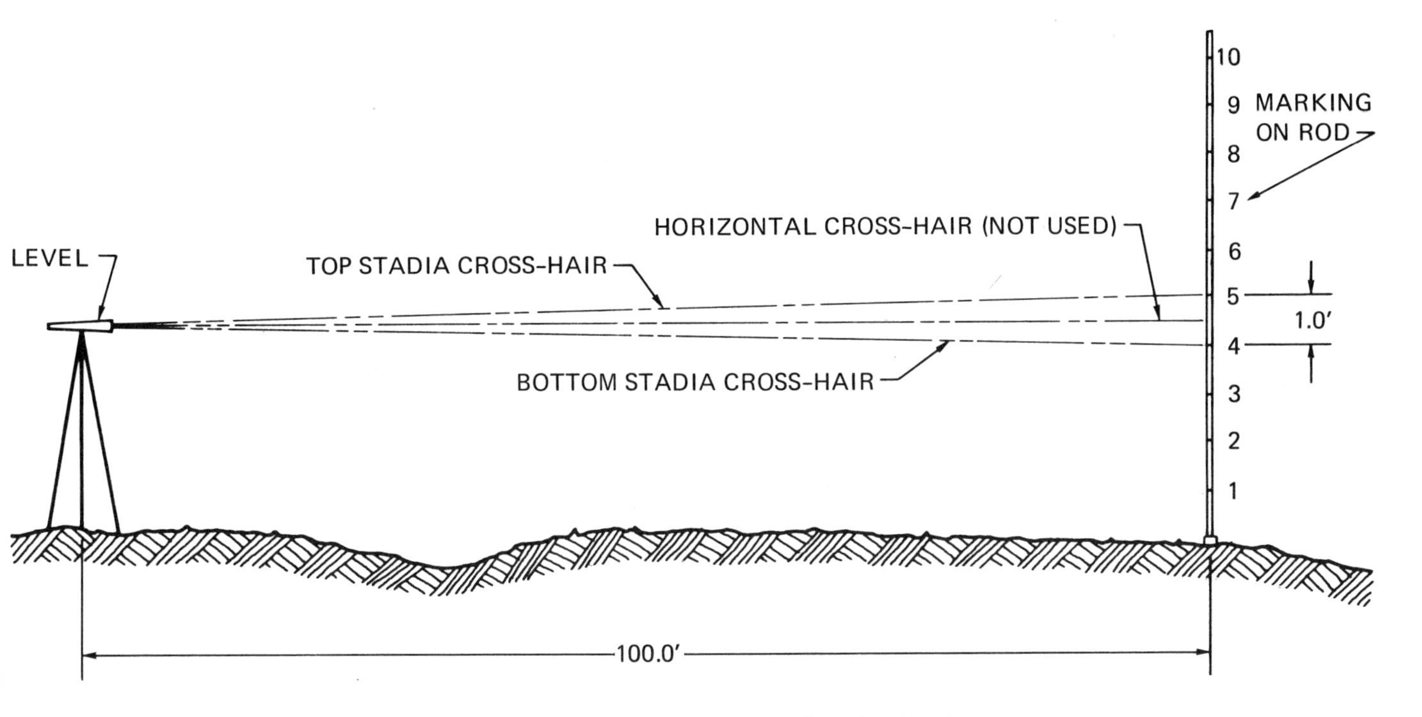

Fig. 3-6 Stadia theory

The level in figure 3-6 is placed 100.0 feet from the rod. The vertical distance between rod marking four and five is 1 foot, so a 100.0-foot horizontal distance measures exactly 1.0 foot on the rod. Example: A reading of 2.5 feet between stadia cross hairs means 250.0 feet from transit to rod.

This is an efficient way to obtain distances. It is not 100 percent correct, but it is close enough for surveys of a low precision, for topographic work, and general location of points.

LEVELING

Leveling is a very simple process. The only equipment needed is a level (transit), tripod, and leveling rod, figure 3-7. The level, as its name implies, is level at all times and simply transfers a height or reading from one place to another.

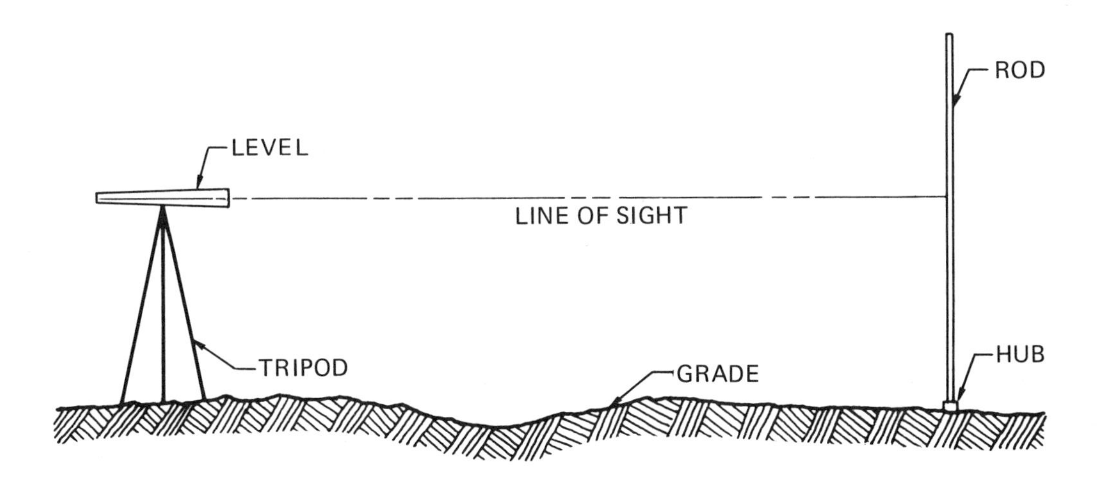

Fig. 3-7 Sighting a level

There are basic terms associated with the leveling process:

Bench mark	–	B.M.
Temporary bench mark	–	T.B.M.
Height of instrument	–	H.I.
Backsight	–	B.S.
Foresight	–	F.S.
Turning point	–	T.P
Station point	–	STA. PT.
Elevation	–	ELEV.

Learn the language of the trade. Memorize these terms and abbreviations. They will be helpful in interpreting field notes.

READING THE ROD

Figure 3-8 shows a Philadelphia rod. It is seven feet long but extends to thirteen feet. It is graduated according to rigid specifications. Notice how the even foot and the .5 graduation is emphasized by a spur-extended graduation. Study how the sample reading is taken in figure 3-8.

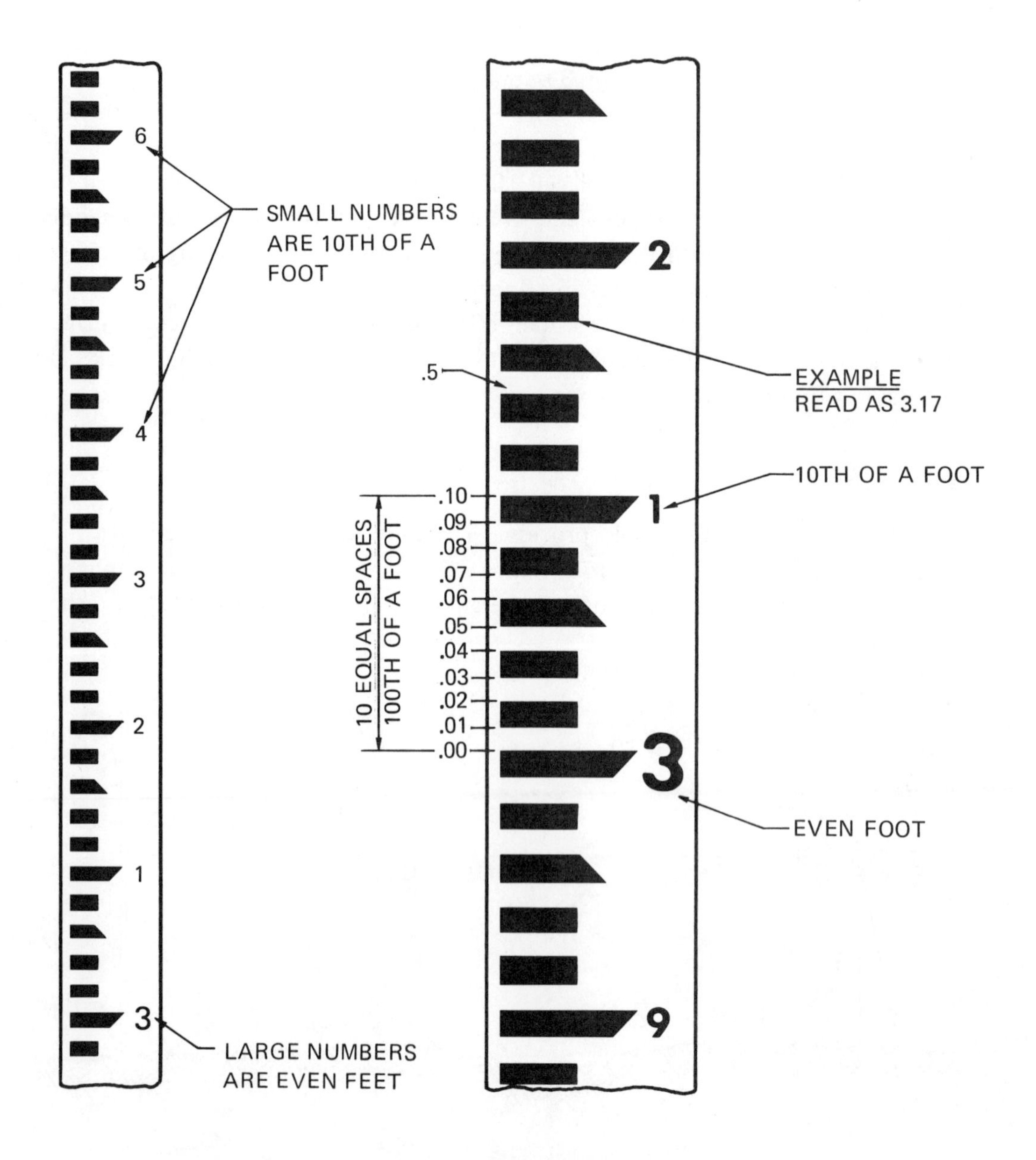

Fig. 3-8 Reading a rod

Practice Exercise 3-1

Determine the following rod readings to the nearest hundredth of a foot. Record the answers in the space provided. Compare your work to the answers on page 70.

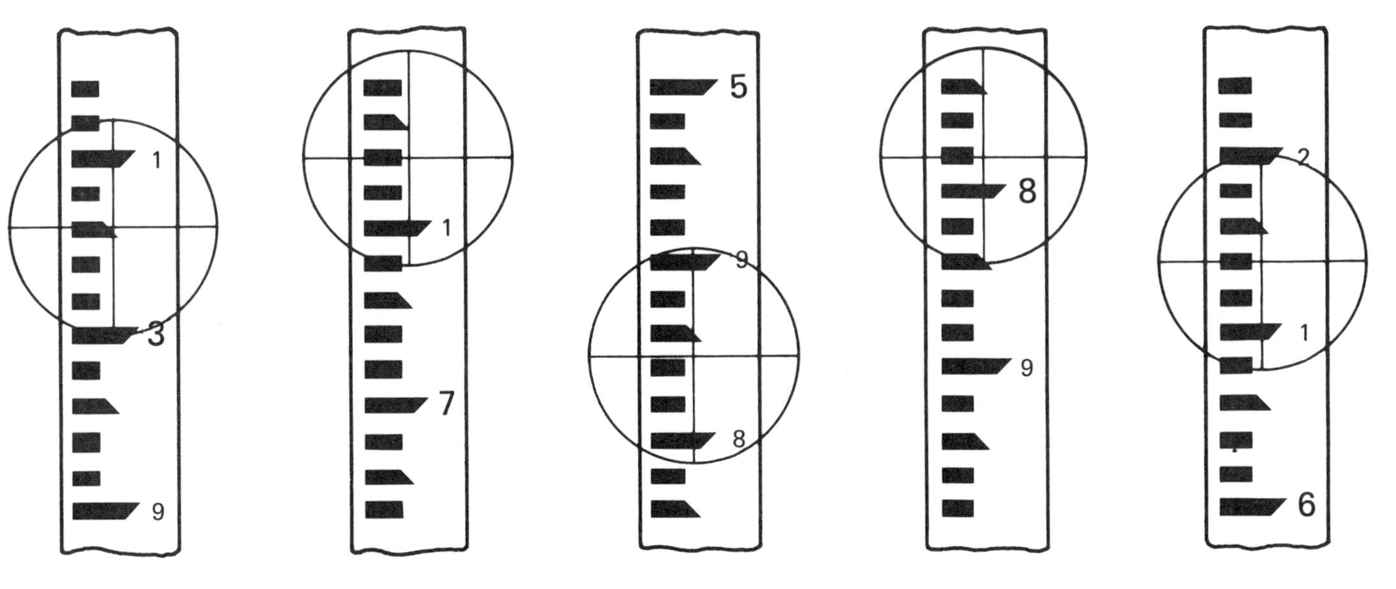

INDICATE THE ABOVE ROD READINGS BELOW:

1.________ 2.________ 3.________ 4.________ 5.________

Holding a Rod Plumb

When leveling, the rod must be held *plumb,* or vertically straight, in order to obtain a correct reading. Study figure 3-9.

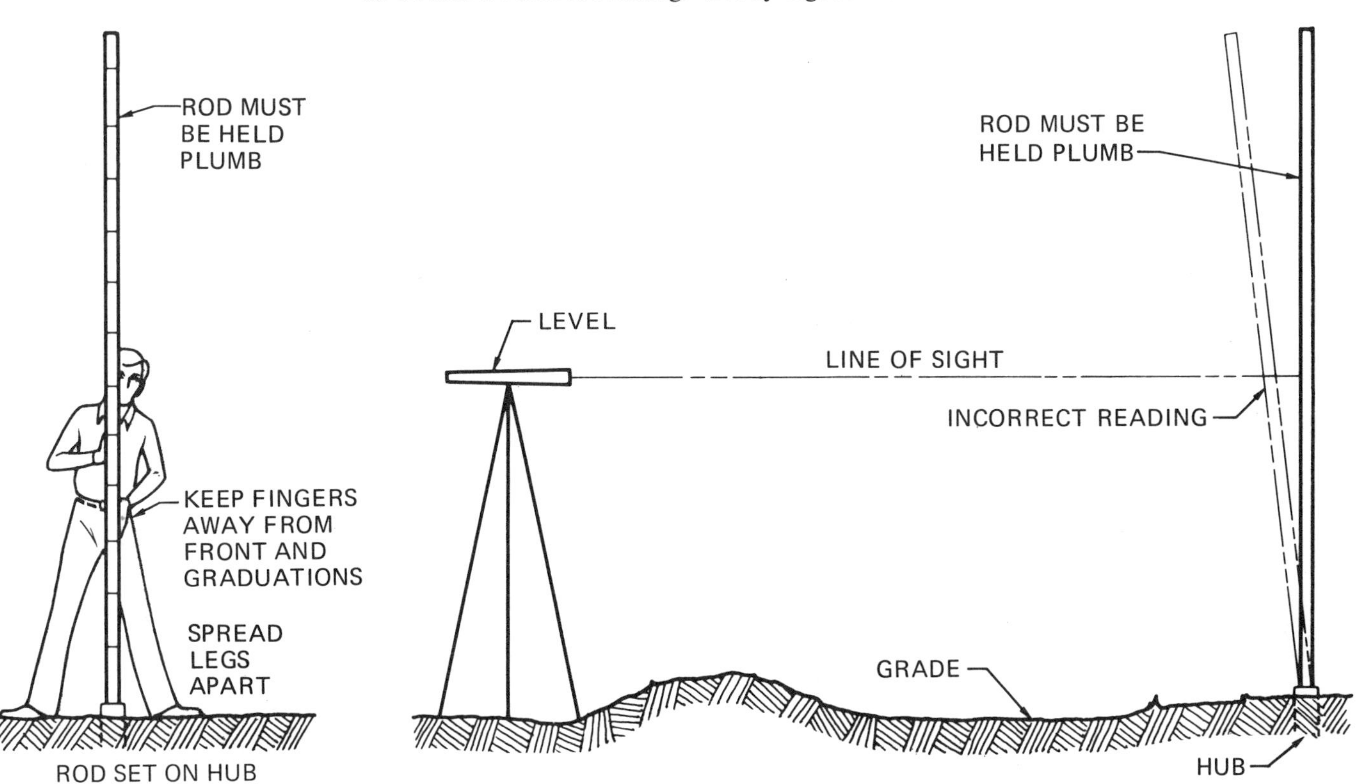

Fig. 3-9 Holding a rod plumb

Field Notes

Figure 3-10 is a sample page from the log that a survey recorder keeps. It is used to record data that is calculated in the field.

STA.	B.S. (+) BACKSIGHT	H.I.	F.S. (−) FORESIGHT	ELEV.
B.M. 1	7.11 =	728.16		721.05
T.P. 1	8.83	735.75	1.24 =	726.92
T.P. 2	11.72	746.36	1.11	734.64
B.M. 2	4.32	740.47	10.21	736.15
T.P. 3	3.06	733.57	9.96	730.51
T.P. 4	2.74	727.40	8.91	724.66
T.P. 5	0.81	716.59	11.62	715.78
B.M. 3			12.42	704.17
Σ B.S. =	38.59	Σ FS =	55.47	
			38.59	
		DIFF =	16.88	CK

(ELEVATION OF H.I. SHOULD ALWAYS BE RECORDED)

SURVEY PARTY / DATE	WEATHER
U.S.G.S. BM ETC.	
CURB	
SPIKE IN A POLE	
CONCRETE MONUMENT IN STREET CORNER	

THIS PAGE SHOULD BE RESERVED FOR REMARKS AND BM DESCRIPTION.

Fig. 3-10 Field note page sample

FINDING UNKNOWN ELEVATIONS

To find an unknown elevation, start from the known elevation at the bench mark and set up the level as illustrated in figure 3-11. Take a reading to find the height of instrument (H.I.). Add it to the known elevation found at the bench mark. Set the rod on the unknown elevation, take a reading, and subtract from the height of instrument. This is the *unknown elevation.*

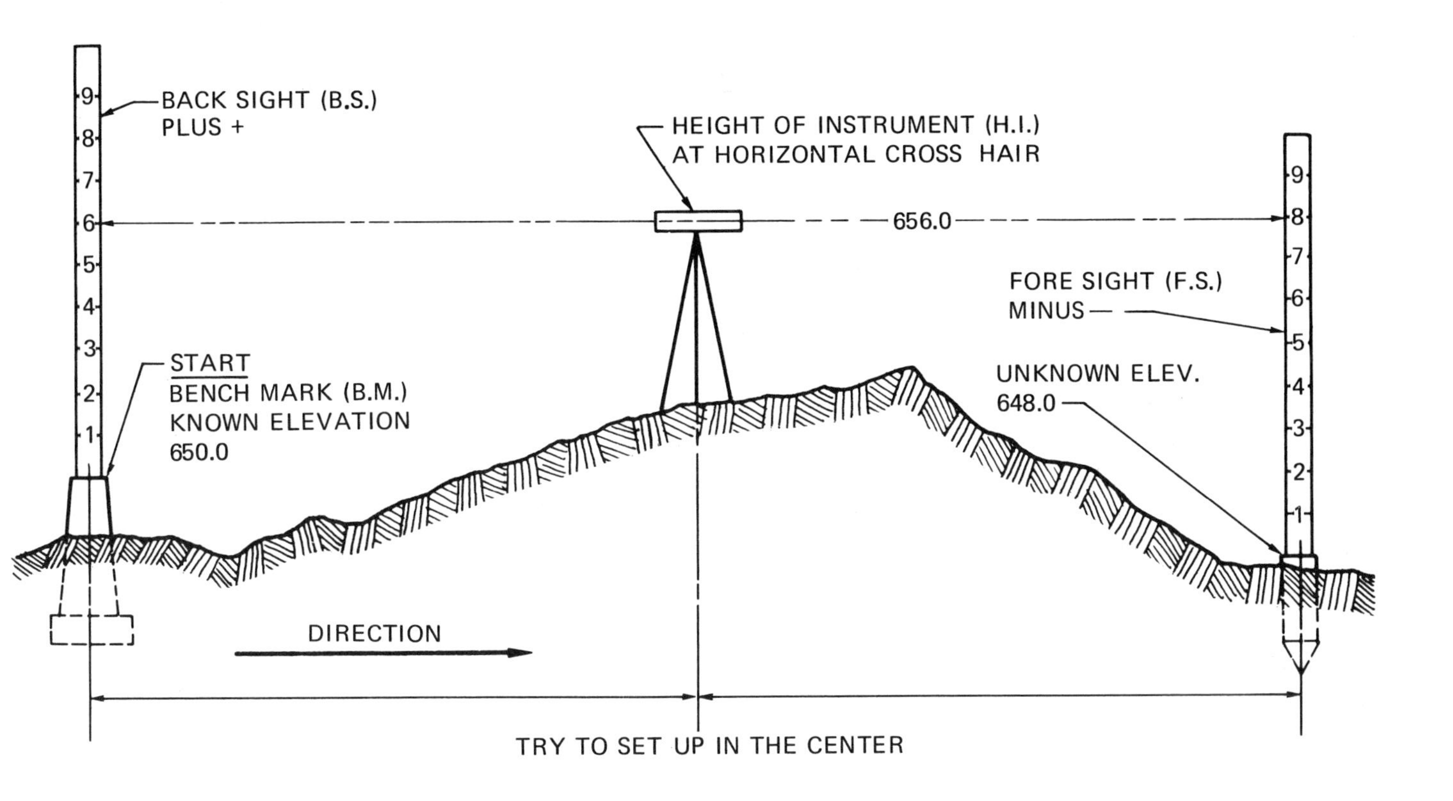

Fig. 3-11 Finding the unknown elevation

Figure 3-12 shows how to record this information in the log. Be sure you understand each entry and what information goes into each column.

STA.	(+) BACKSIGHT	H.I.	(–) FORESIGHT	ELEV.
B. M.				650.0
	6	656.0	8	648.0

Fig. 3-12 Sample of field notes

TURNING POINTS

Turning points are used for sighting over long distances or sights that cannot be seen. A turning point (T.P.) is nothing more than a temporary bench mark (T.B.M.). Take one reading as a foresight (–), reset the level, and use the same point as a backsight (+), figure 3-13.

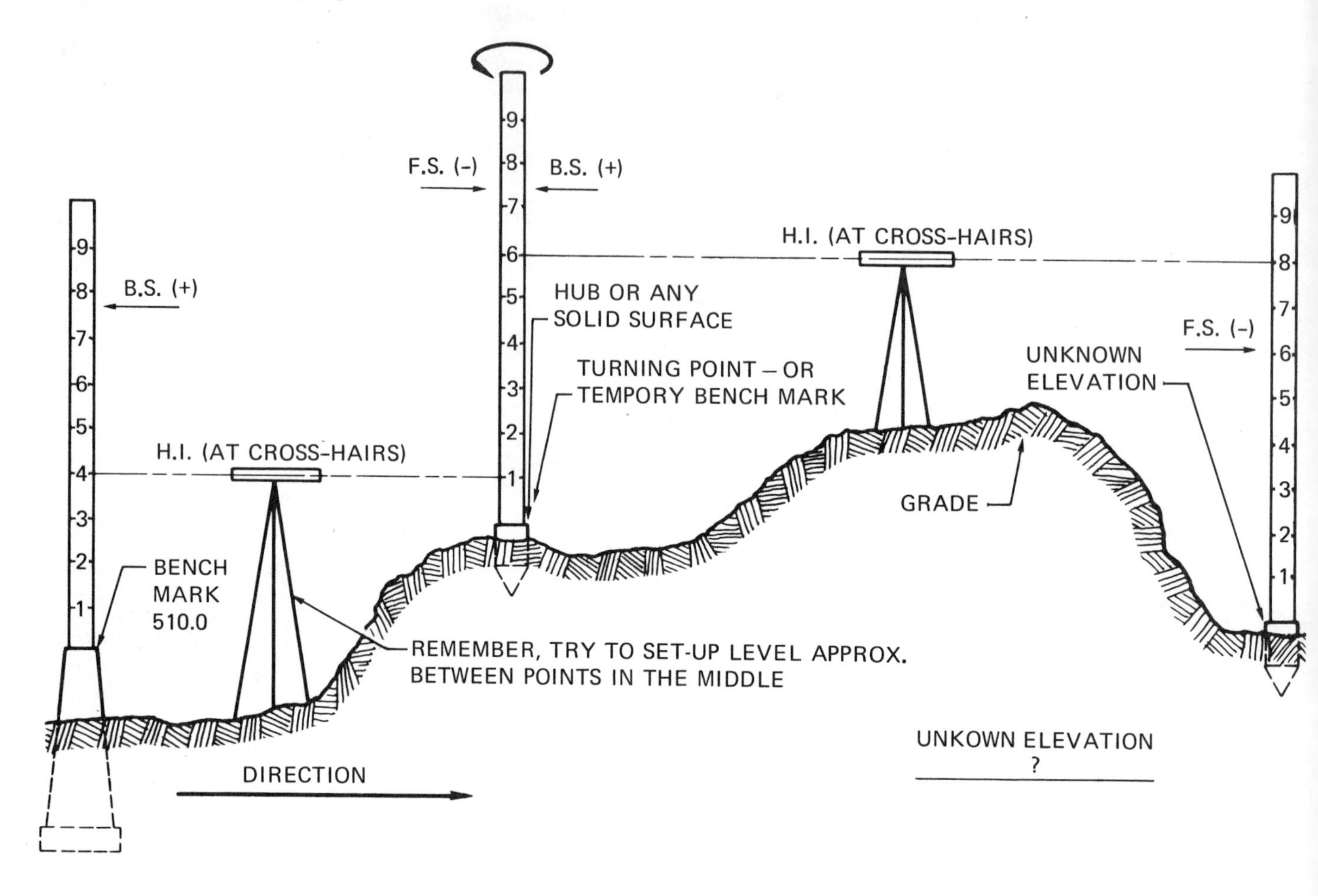

Fig. 3-13

These measurements are recorded in figure 3-14.

STA.	(+) BACKSIGHT	H.I.	(–) FORESIGHT	ELEV.
B. M.	–	–	–	510.0′
	4	514.0	1	513.0′
T. P. 1	6	519.0	8	511.0′

Fig. 3-14 Sample of field notes

Many times several turning points must be made, figure 3-15. A turning point can be a hub set in the ground, an edge of a sidewalk, a large stone in the field, a nail set in a tree, a side of a ledge, etc., all of which are clearly marked with chalk or paint. Many T.B.M.s are seen along new construction sites.

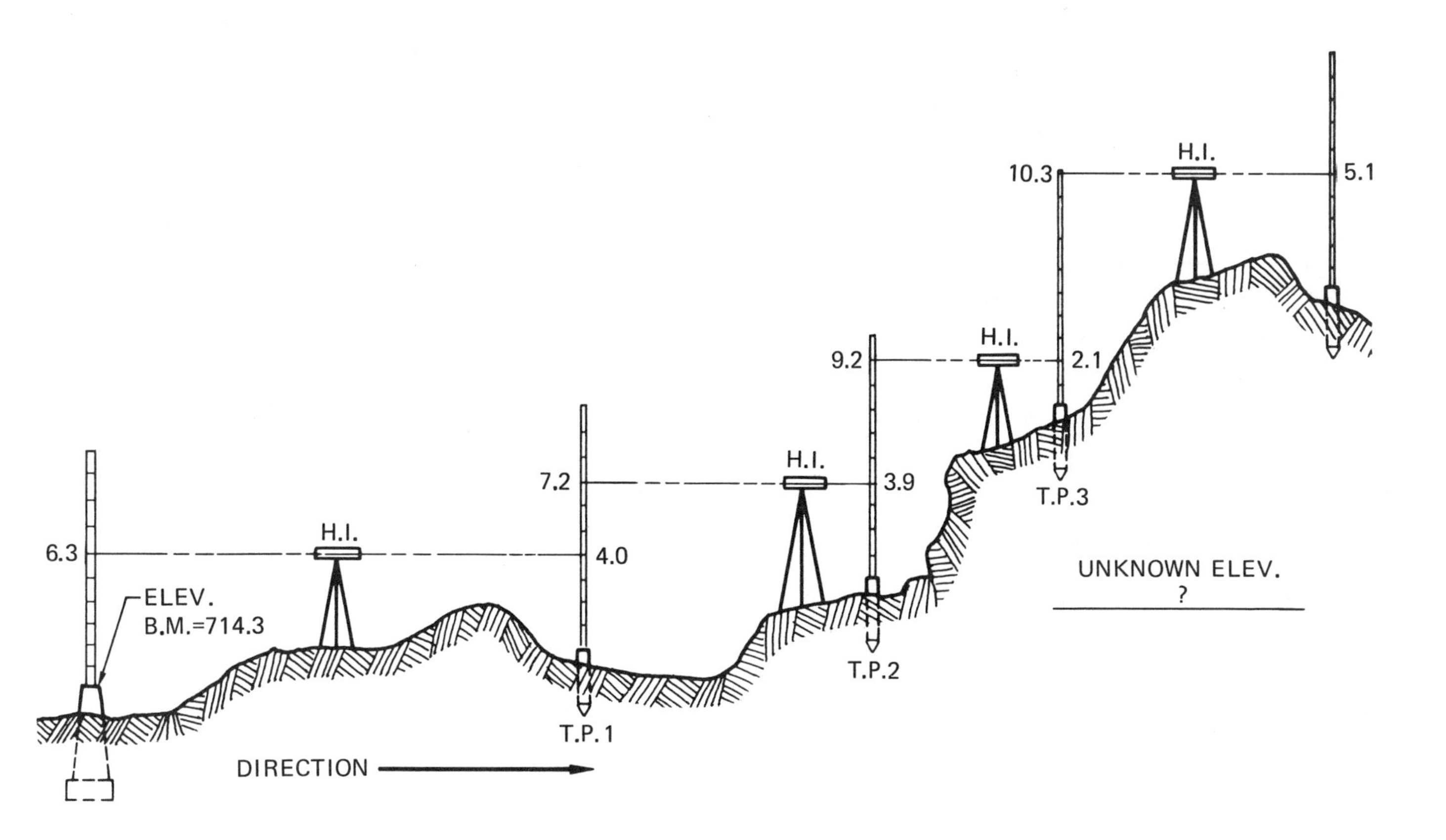

Fig. 3-15 Measuring, using several turning points

Practice Exercise 3-2

Using the example in figure 3-15, complete the following field note page. Compare your work to the answer on page 70.

STA.	(+) BACKSIGHT	H.I.	(–) FORESIGHT	ELEV.
B.M.	–	–	–	714.3
	6.3	720.6	4.0	716.6
T.P.1	7.2			

Practice Exercise 3-3

Fill in all of the empty spaces in the field note page given. Compare your work to the answer on page 70.

STA.	(+) BACKSIGHT	H.I.	(–) FORESIGHT	ELEV.
B.M. 1	—	—	—	410.52
	5.17		7.22	
T.P. 1	3.11		6.08	
T.P. 2	5.25		3.61	
B.M. 2		412.74	1.52	
T.P. 1	6.15			410.85
T.P. 2	7.80			405.82
T.P. 3		415.13		(412.03)

AERIAL PHOTOGRAPHY

Aerial photography is used to assist the cartographer in mapping large land areas. A series of pictures are taken from an airplane flying a straight course over the tract to be mapped. Each picture overlaps the one preceding it so that they may be aligned to make up the finished map, called a *photomosaic.* The photomosaic is actually a picture taken of all the photos combined on one surface, figure 3-16.

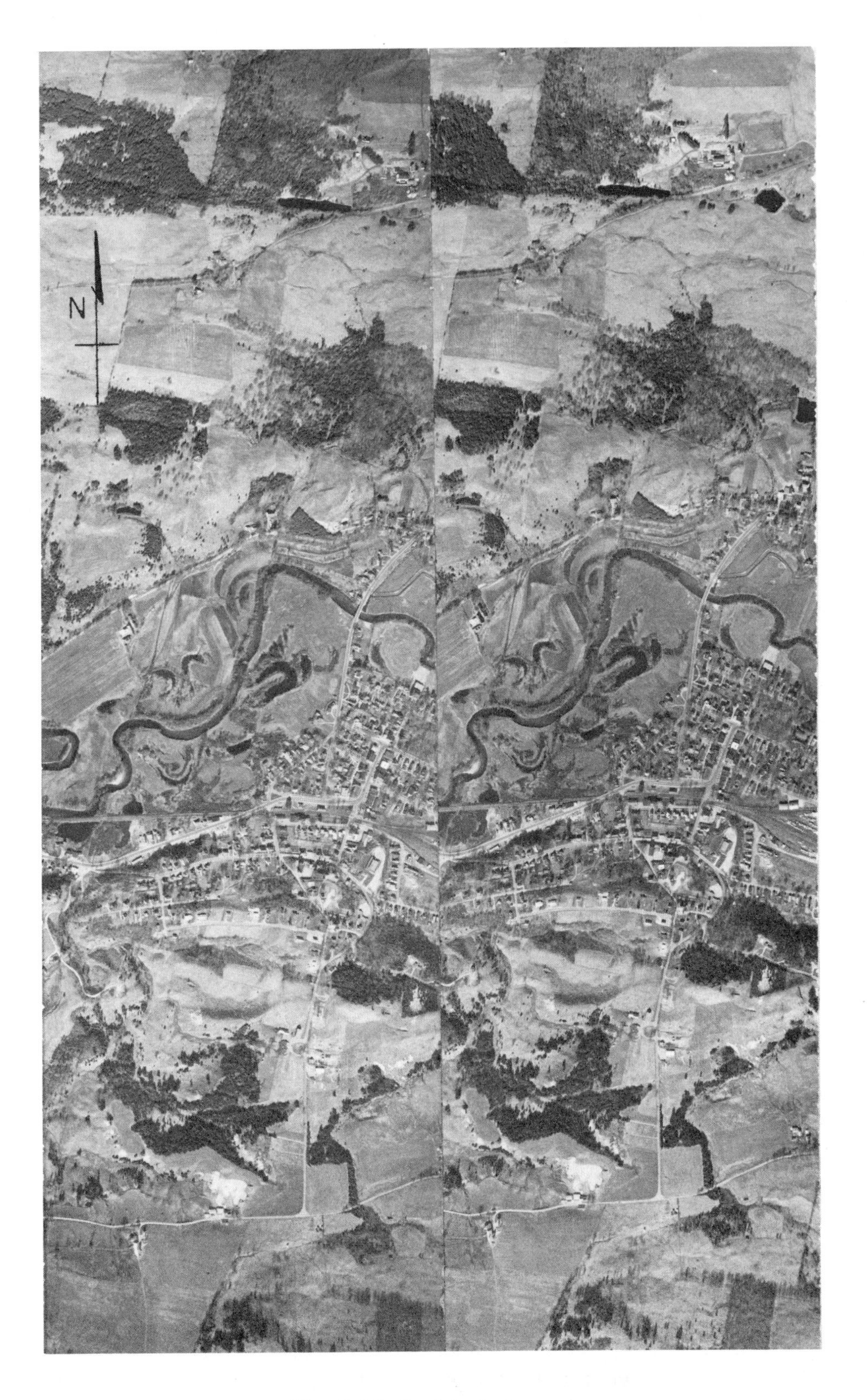

Fig. 3-16 Aerial photograph

Carefully study the aerial photograph. Note all the high and low elevations. Aerial photos are revised every twenty years.

An airplane flies over an area and takes a series of pictures. Each of these pictures overlap about 60 percent, figure 3-17.

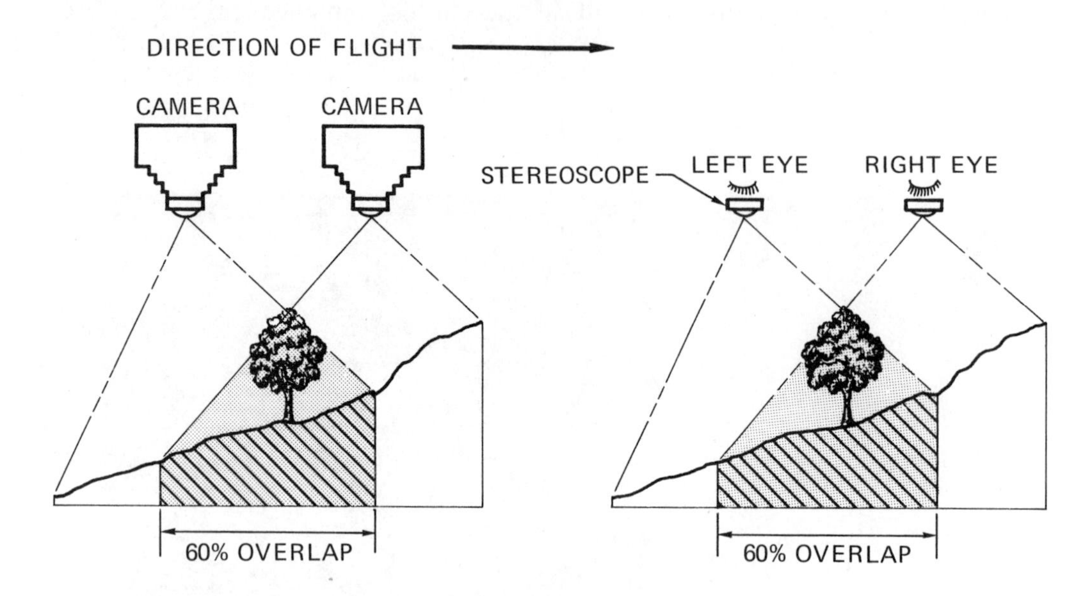

Fig. 3-17

Each photo viewed alone appears flat; but if the two photos are viewed through a stereoscope, they give a full, three-dimensional view as would be seen by actually flying over the area. The *stereoscope* is a devise that enables a cartographer to get a three-dimensional effect when looking at a pair of stereo aerial photographs, figures 3-18 and 3-19.

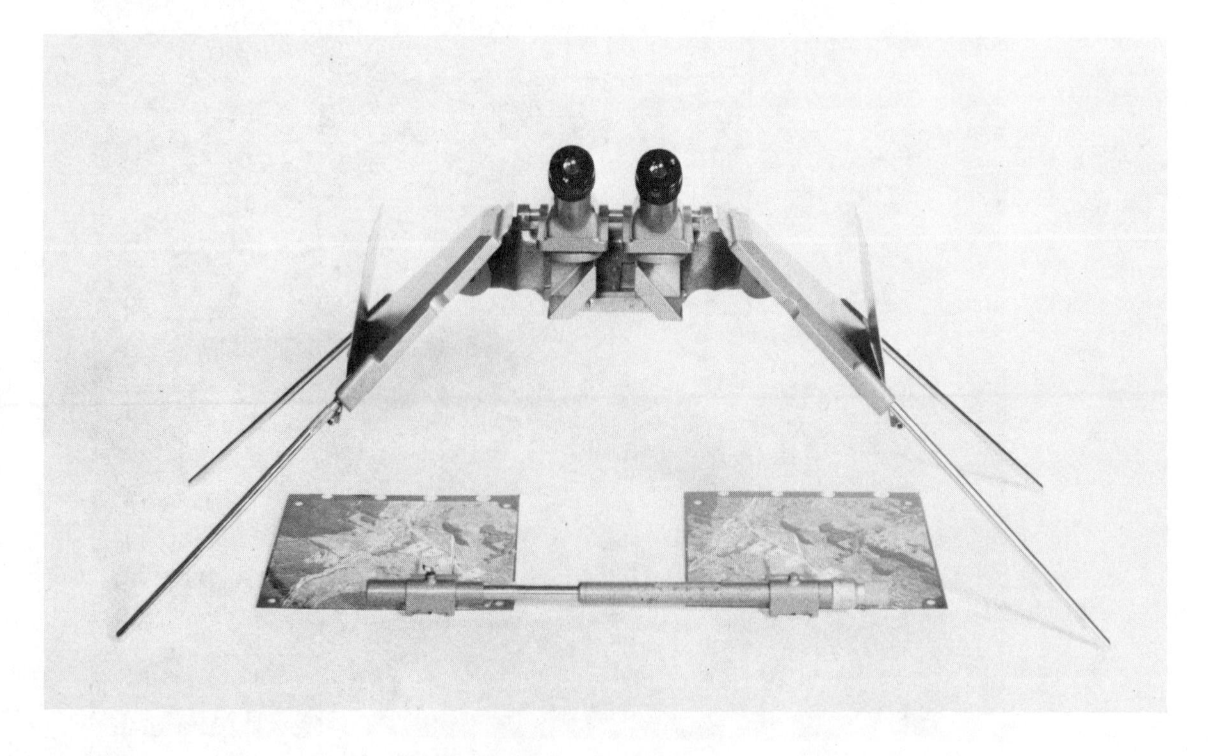

Fig. 3-18 A stereoscope is used to look at a stereo pair of aerial photographs in order to obtain a three-dimensional view of the area.

Fig. 3-19 Stereo pair of aerial photographs

ANSWERS TO PRACTICE EXERCISES

Carefully compare your work to the answer for each exercise. Refer any questions to the instructor.

Exercise 3-1

1. 3.05
2. 7.13
3. 4.84
4. 8.015
5. 6.135

Exercise 3-2

STA.	(+) BACKSIGHT	H.I.	(–) FORESIGHT	ELEV.
B.M.	—	—	—	714.3
	6.3	720.6	4.0	716.6
T.P. 1	7.2	723.8	3.9	719.9
T.P. 2	9.2	729.1	2.1	727.0
T.P. 3	10.3	737.3	5.1	732.2
				UNKNOWN ELEV.

Exercise 3-3

STA.	(+) BACKSIGHT	H.I.	(–) FORESIGHT	ELEV.
B.M. 1	—	—	—	410.52
	5.17	415.69	7.22	408.47
T.P. 1	3.11	411.58	6.08	405.50
T.P. 2	5.25	410.75	3.61	407.14
B.M. 2	5.60	412.74	1.52	411.22
T.P. 1	6.15	417.37	6.52	410.85
T.P. 2	7.80	418.65	12.83	405.82
T.P. 3	9.31	415.13	3.10	412.03

UNIT REVIEW

45-minute time limit

Problem 1

Determine the following rod readings to the nearest hundredth of a foot. Record the answers in the spaces provided.

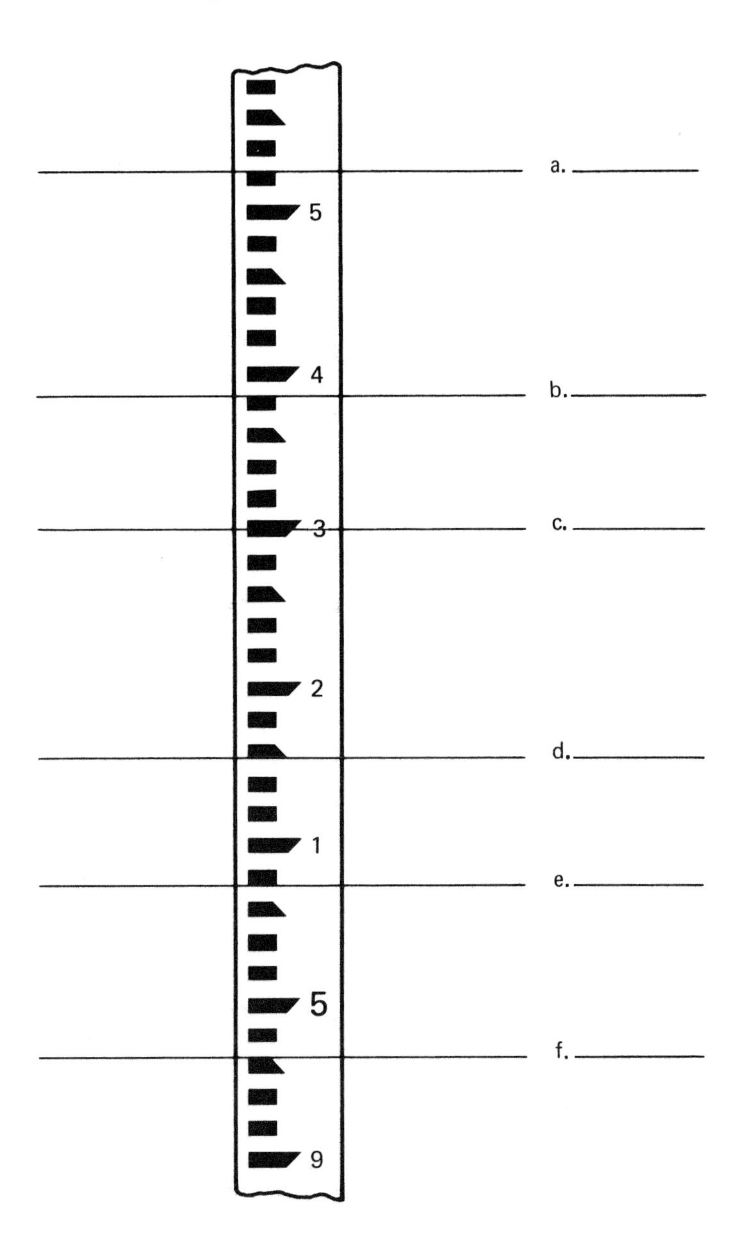

Problem 2

A. Find the unknown elevation and place the answer in the space provided. Fill in log as you calculate answer.

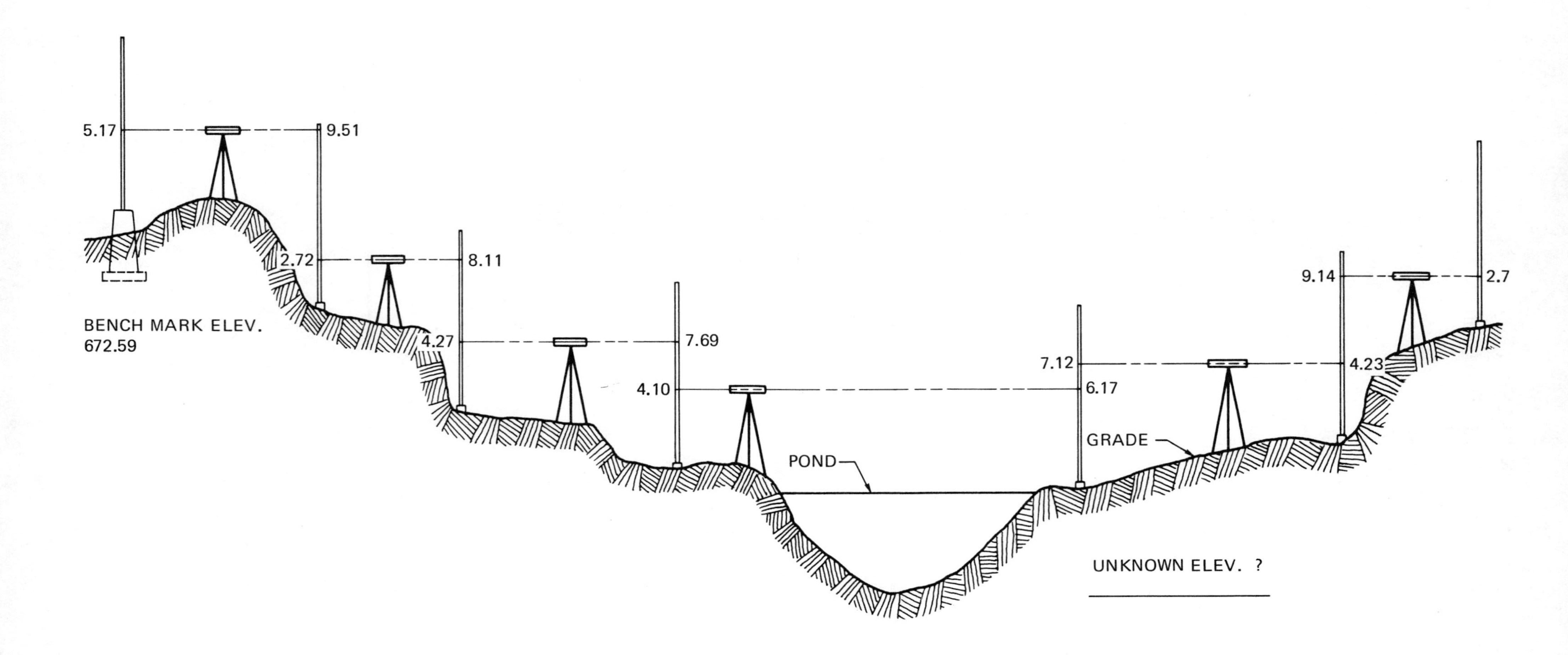

B. Record the field notes necessary to determine the unknown elevation.

STA.	(+) BACKSIGHT	H.I.	(–) FORESIGHT	ELEV.

Problem 3

Using a stereoscope lens, locate and outline the lowest area on the aerial photo. Indicate it by an "L." Then locate and outline five or six high areas, indicating them by an "H."

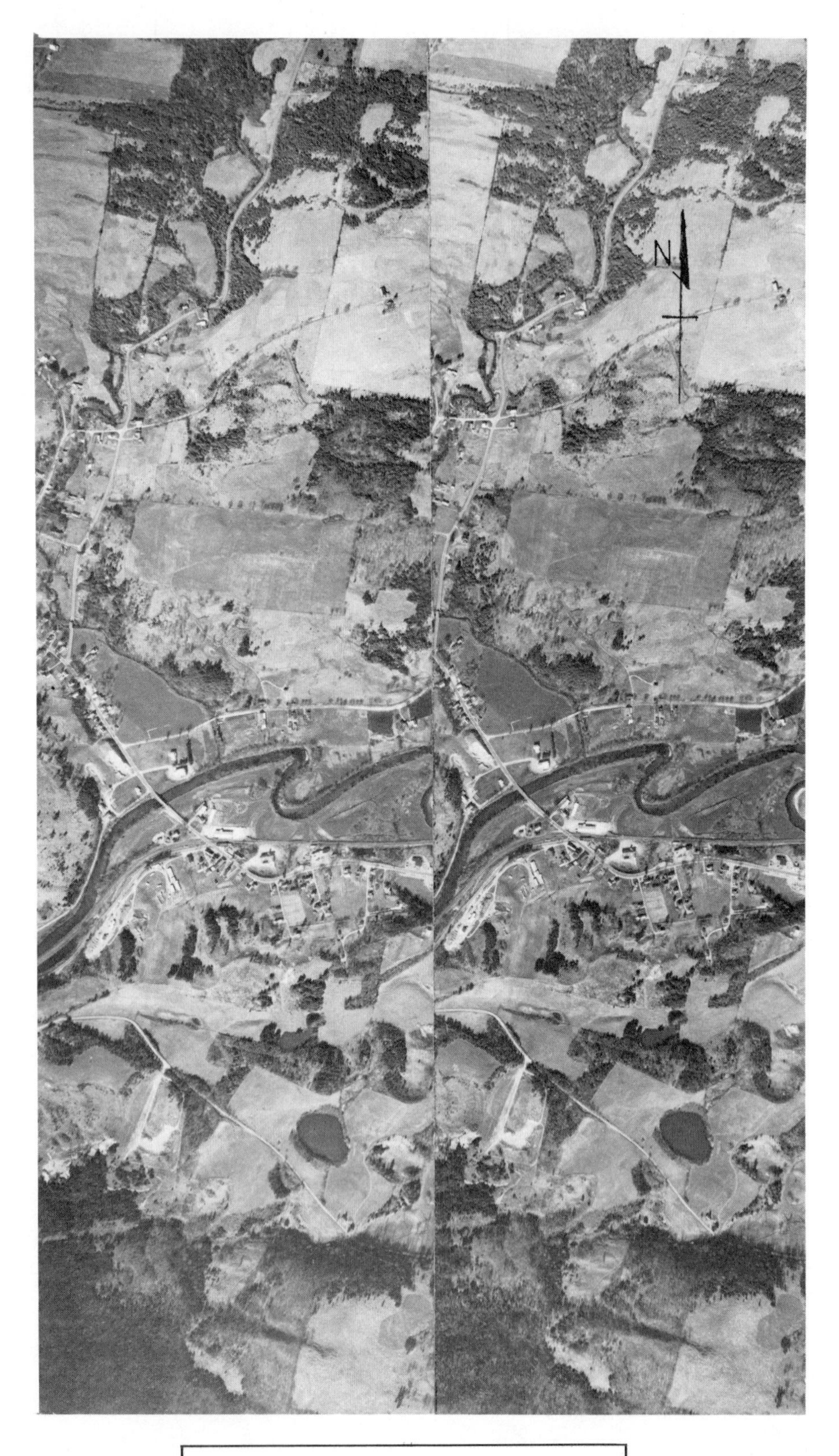

Before proceeding to the next unit:

_______ Instructor's approval

_______ Progress plotted

UNIT 4

CONTOUR MAPS

OBJECTIVE

The student will learn how to draw contour maps from field notes.

PRETEST

One-hour time limit

Construct a contour map from the field notes given. Use standard drafting methods and line weight.

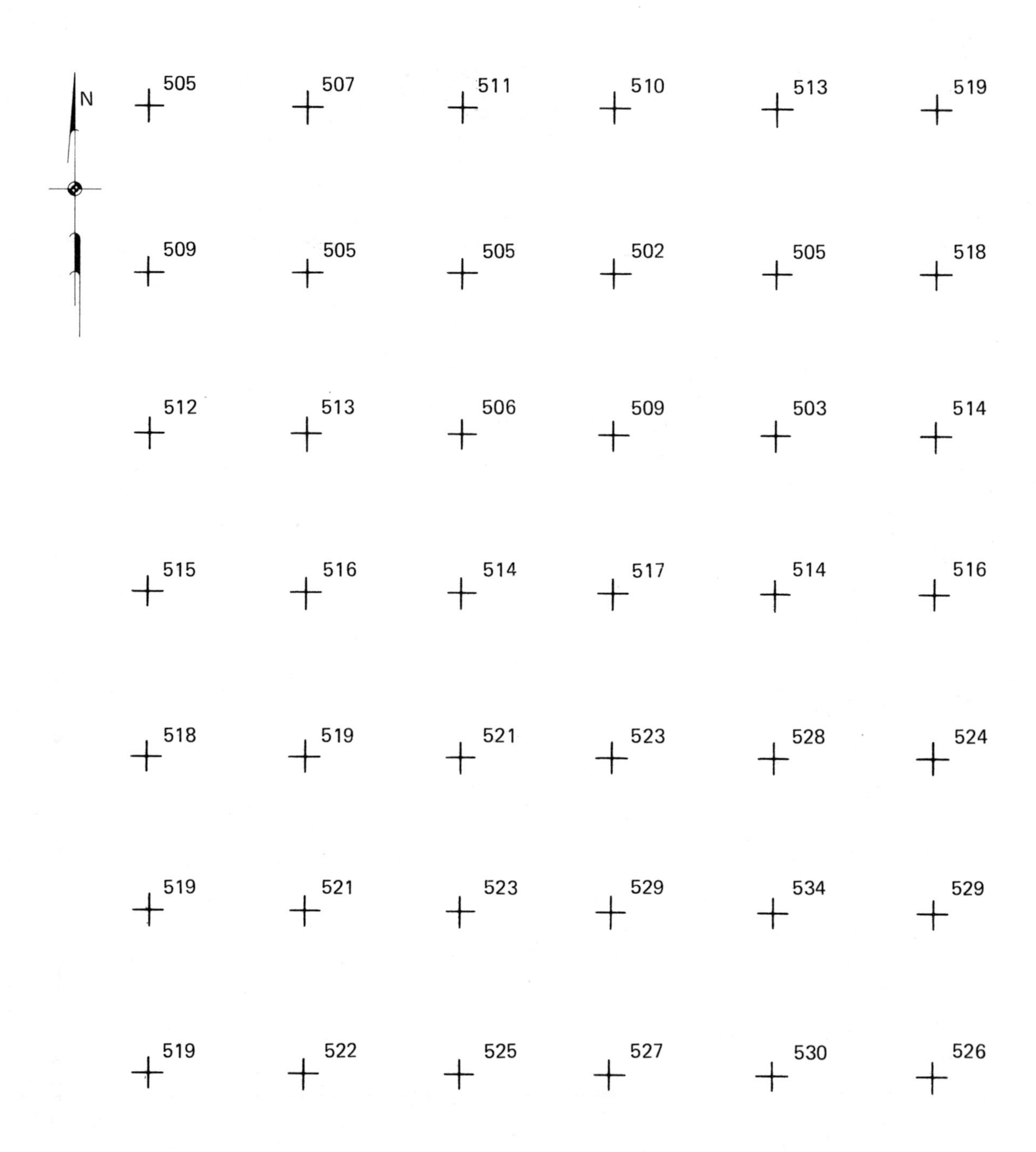

RELATED TERMS

Give a brief definition of each term as progress is made through the unit.

Contour map ____________________

Sea level ____________________

Datum line ____________________

Contour line ____________________

Index ____________________

Intermediate line ____________________

Supplemental line ____________________

Depression ____________________

Spot elevation ____________________

Profile ____________________

Equal-spacing divider ____________________

Station point ____________________

Colored flag ____________________

Marking stake ____________________

Hub ____________________

Cut ____________________

Fill ____________________

Slope ____________________

CONTOUR MAPS

Before a topographic map can be drawn, the contour of the land must be determined. A *contour map* shows the elevation of the land surfaces and the depth of water features. Once the contours are drawn, other natural and man-made features are added to complete the topographic map.

CONTOUR LINES

Elevations on a topographic map start at "0," which is *sea level* between high and low tides. Sea level is the *datum line* so every point on earth is measured above or below sea level.

A *contour line* is an imaginary line on the ground connecting all points that are the same elevation above or below sea level. They show the shape of the ground at each elevation. When contour lines are close together, the ground slopes very steeply. When they are spread out, the ground slopes moderately.

There are various kinds of contour lines, figure 4-1. They vary in line thickness. The elevation of the contour is numbered in a broken-out space on the line.

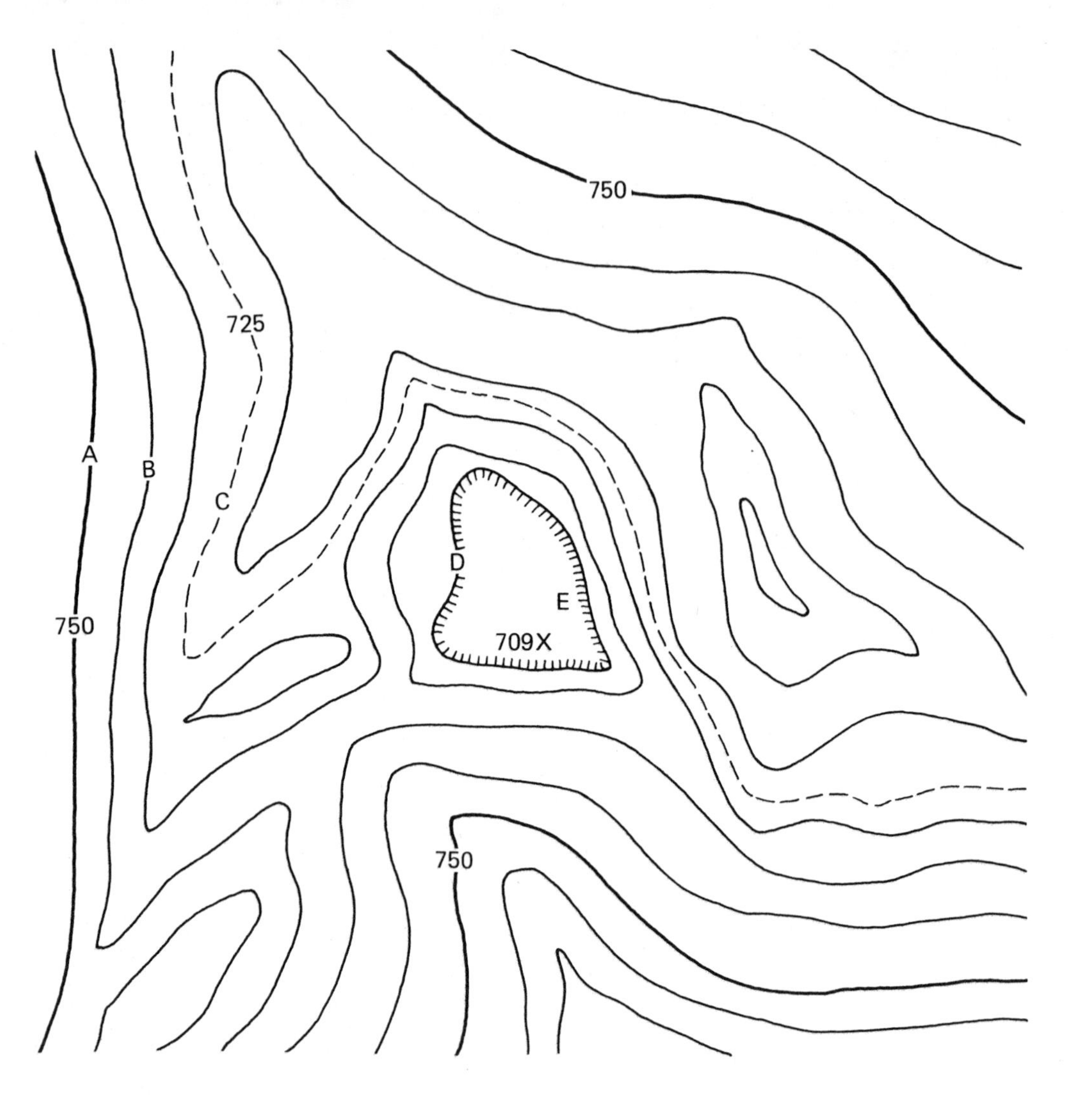

A INDEX–
Every 5th line, numbered
B INTERMEDIATE–
Between index contours
C SUPPLEMENTAL–
Represents half interval
D DEPRESSIONS–
Much lower elevation
E SPOT ELEVATION–
Locates an exact elevation at a particular site

NOTE LINE THICKNESS

Fig. 4-1 Contour lines

CONTROL SURVEYS

Control surveys are required to present map features in correct relationship to each other and to the earth's surface. Vertical control provides the correct location for the contour lines. Permanent control points, often called *bench marks,* are metal markers 3 3/4 inches in diameter, figure 4-2, set in cement and anchored three to six feet into the ground. They are shown on maps by appropriate symbols.

Control points indicate the elevation of the point marked and the date the measurement was taken.

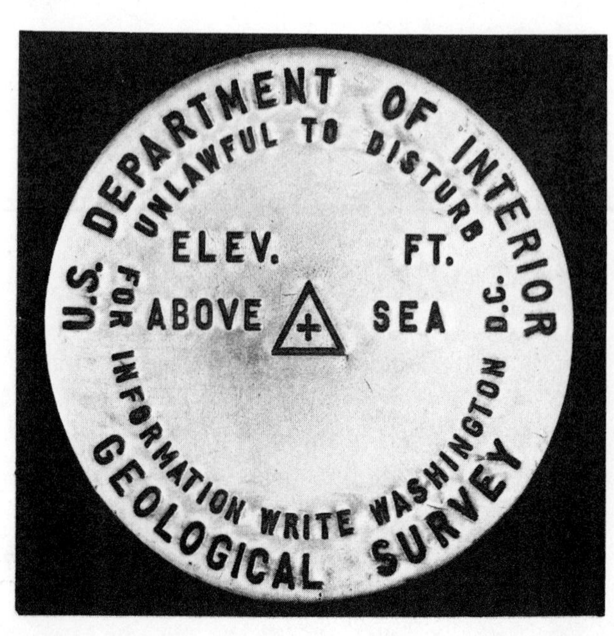

Fig. 4-2 Control point

PROFILE VIEW

A *profile view* is a vertical plane through a contour map that shows the vertical outline of the surface. Figure 4-3 shows a contour map of an area with a cutting-plane line (A-A) passing through it. Notice points A, J, and S. Each point is projected down the same way to form the profile or section view. The grid at the left side indicates the elevation in feet above sea level.

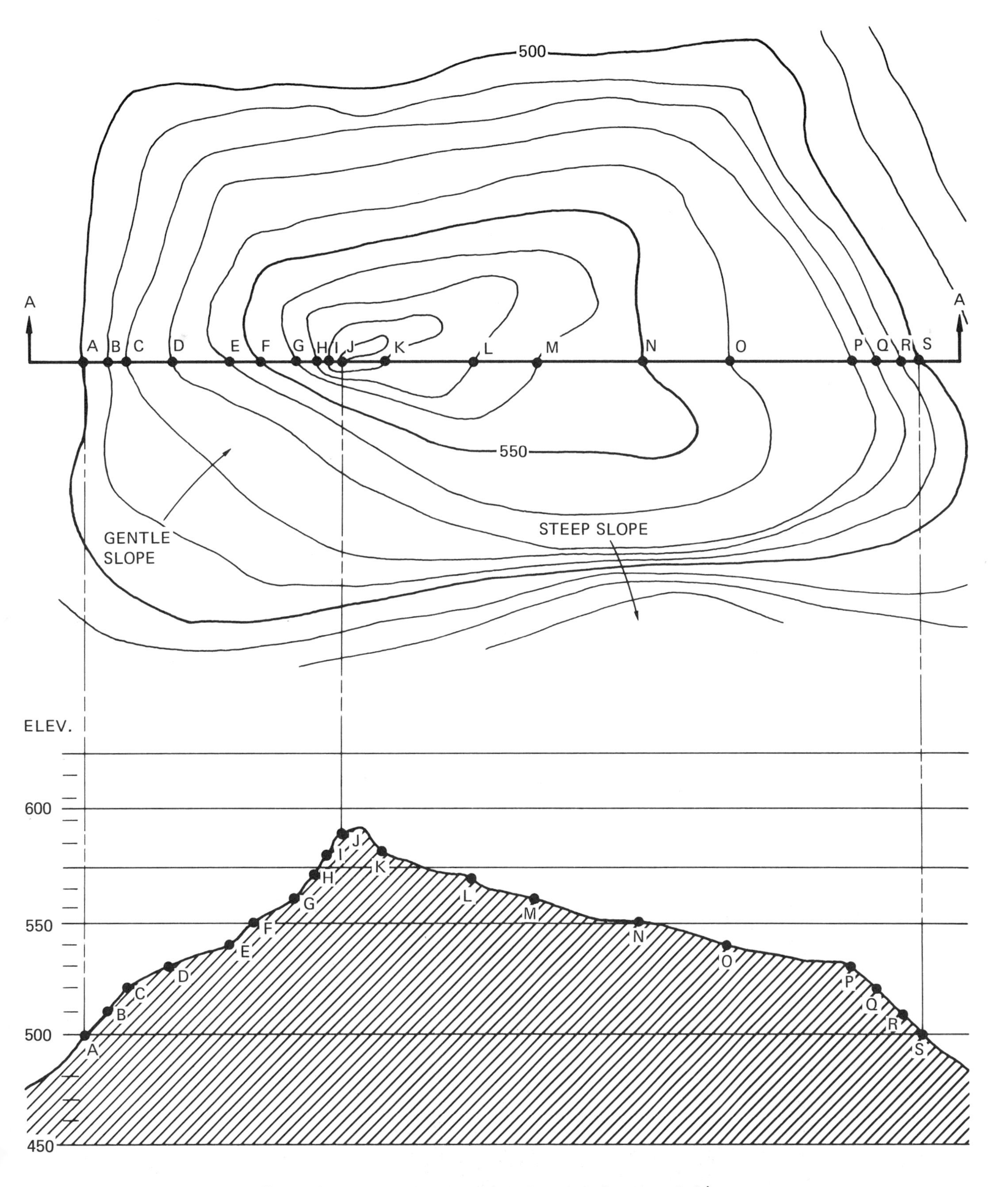

Fig. 4-3 Contour map with profile view (section A-A)

CONTOUR LAYOUT

A contour map drawn from field notes is not 100 percent correct. Sometimes a drafter must visit the site in order to complete a drawing or check the finished product. An experienced drafter, however, can come very close the the actual land shape by a method that, in theory, is correct.

In figure 4-4, point A is 615 feet, point B is 623 feet, and point C is 612 feet above sea level. It is, therefore, a good assumption that the ground slope is constant from A to B and from B to C. If a line is drawn from A to B to C (dashed), one has a fairly accurate actual ground illustration.

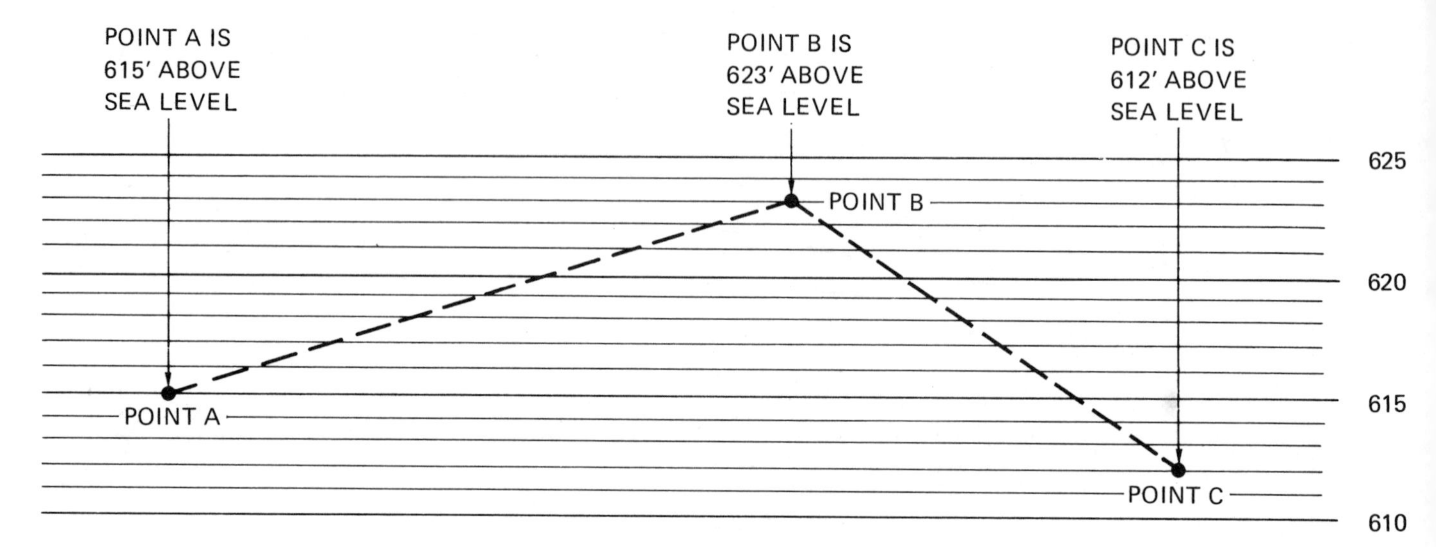

Fig. 4-4

A profile view of figure 4-4 looks like figure 4-5. Notice that from A to B there are, in theory, even spaces from 615 feet to 623 feet, and from B to C there are even spaces from 623 feet to 612 feet. A visit to the site would reveal any large dips or hollows that should be added to the illustration. The land could actually have a contour like that indicated by the thin line in the profile view, but it is assumed straight and drawn as indicated by the thick line.

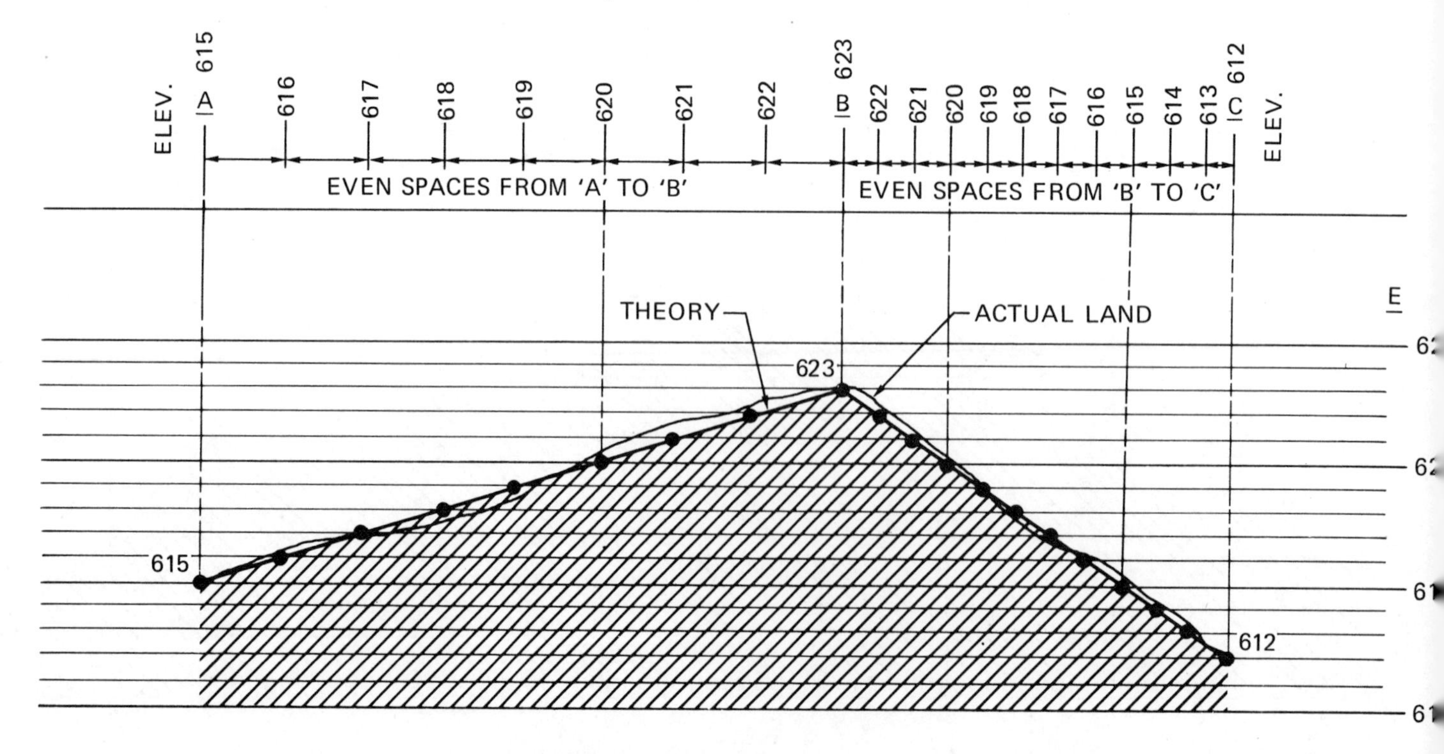

Fig. 4-5 Profile view of figure 4-4

EQUAL-SPACING DIVIDERS

Equal-spacing dividers are simple to use, figure 4-6. It is a very delicate and expensive tool, however, so care must be taken while working with it. As the name implies, the equal-spacing divider divides a given distance into a given number of equal parts.

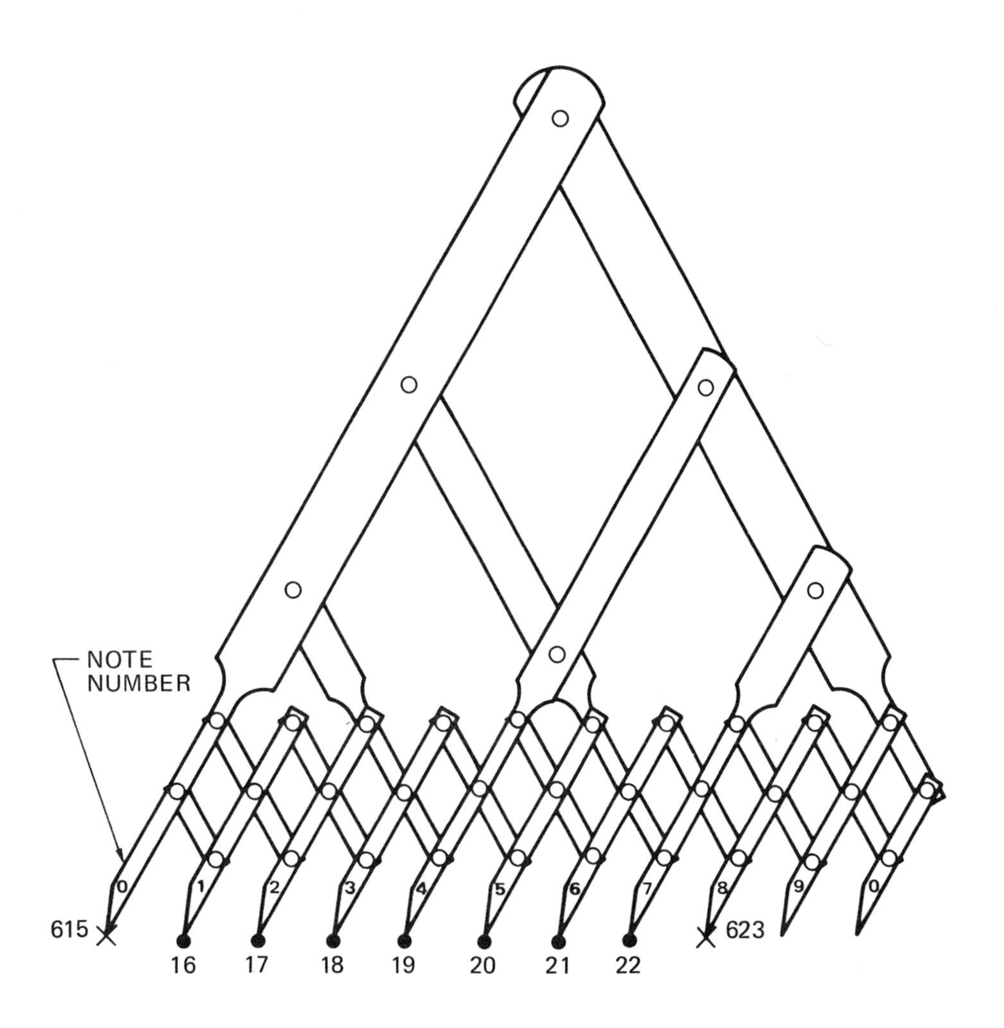

Fig. 4-6 Equal-spacing divider

Notice the numbers 0, 1, 2, 3, 4, 5, 6, 7, 8, 9, 0 at the tip of each point, figure 4-16. Starting with tip #0 on 615 foot elevation, move the points in or out to place point #8 on 623 foot elevation (623 minus 615 = 8). Put a dot at the end of each point and number 16, 17, 18, 19, 20, 21, 22. #9 and #0 at the right end are not necessary. With a little practice one can equally space any area very quickly. Be sure to *number each point* so there is no chance of error.

Using Field Notes to Lay Out a Contour Map

Step 1. Study figure 4-7. This is the information a drafter has to start a layout of a contour map. These figures were made directly in the field by a survey crew. They indicate the exact elevation above sea level at those points, 100 feet apart, as illustrated.

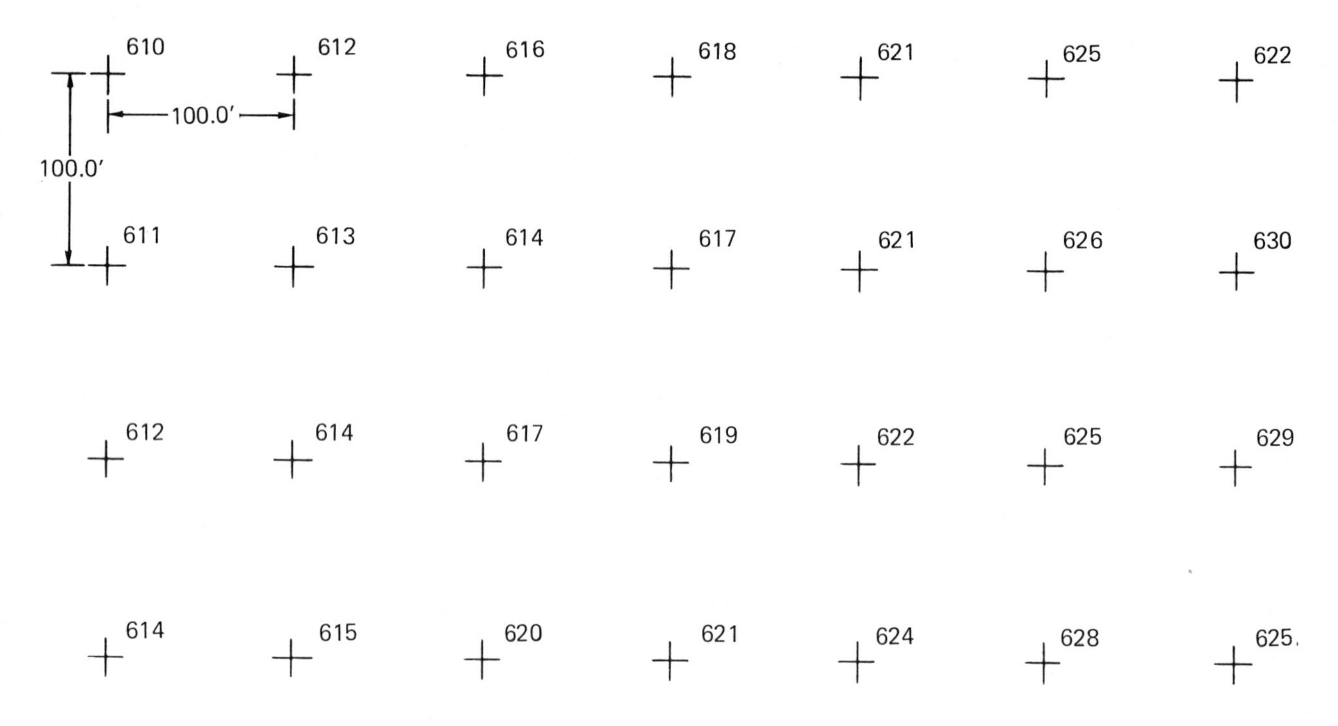

Fig. 4-7 Field notes

Step 2. Starting at the top left corner (610), go to the first point to the right (612), figure 4-8. It can be assumed that 611 feet is at the center as indicated. From 612 the next point is 616 feet. It can be assumed that there are four even spaces shown as 613, 614, and 615. Follow this procedure both up and down and from left to right, using equal-spacing dividers. Locate and label each point.

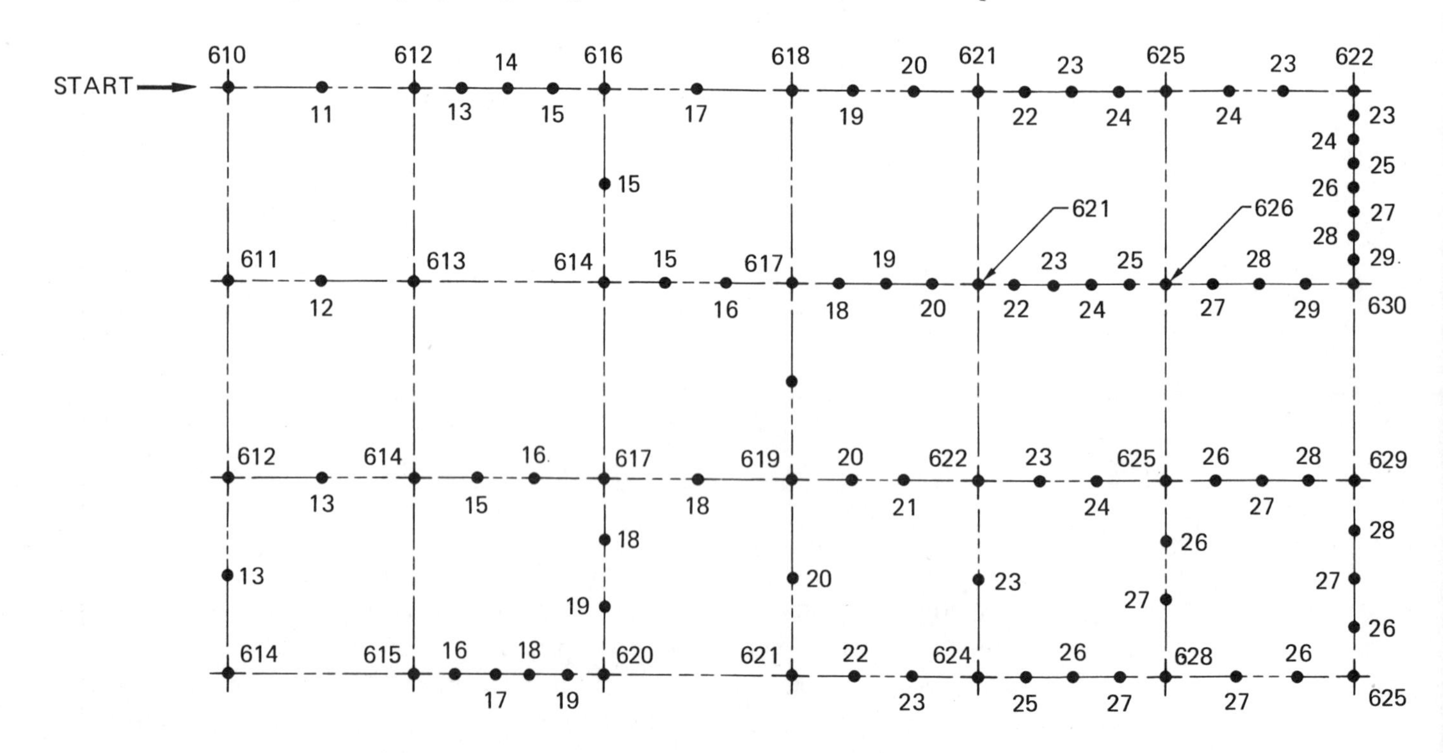

Fig. 4-8 Locating missing elevations

Step 3. Very carefully connect all lines of the same elevation, figure 4-9, with light, straight lines. Lines should not cross. Sometimes there is a question, as indicated by the (?) at elevation 625. The elevation seems to go in two directions. This problem can be solved by reasoning or by a visit to the site to determine what the ground level actually does. See how it was solved in figure 4-10.

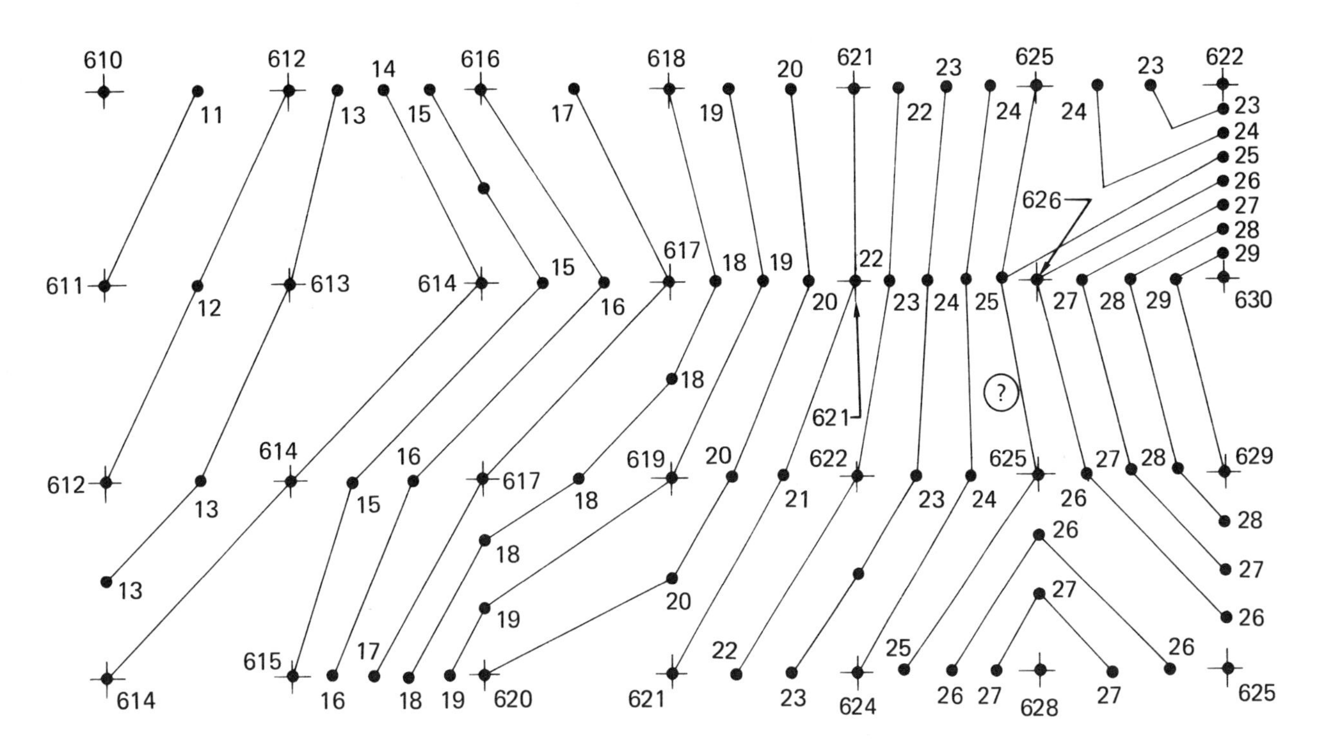

Fig. 4-9 Connecting elevation lines

Step 4. Select the index contour lines as they will be numbered and have a thicker line than the rest, figure 4-10. Because this process is approximate anyway, carefully freehand draw the index contours with a thick line. Draw the intermediate contours next. Round all sharp corners. Many times the contour lines are parallel to each other. If in doubt, make the lines parallel.

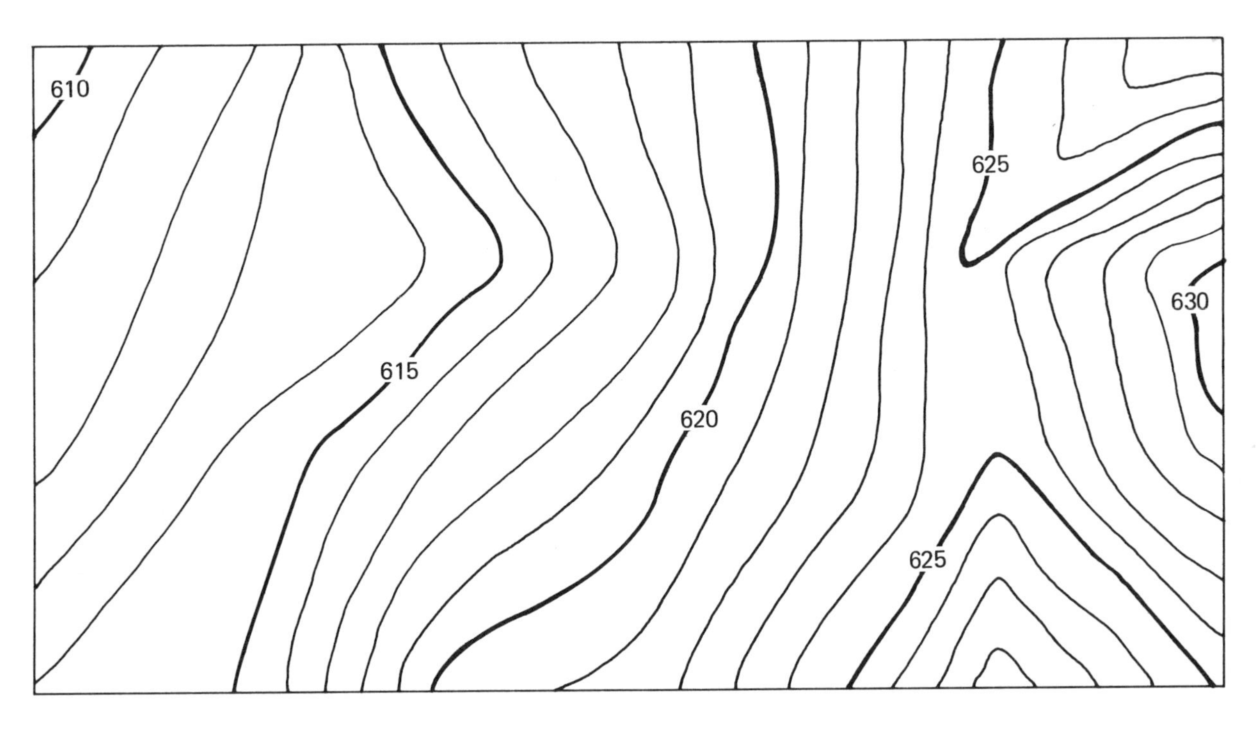

Fig. 4-10 Finished drawing

Practice Exercise 4-1

Draw a contour map from the field notes given. Locate the pond and shade it in. The pond is at an elevation of 707 feet above sea level. Locate and call out the higest point using standard illustrations for a spot elevation. Compare your work to the answer on page 89.

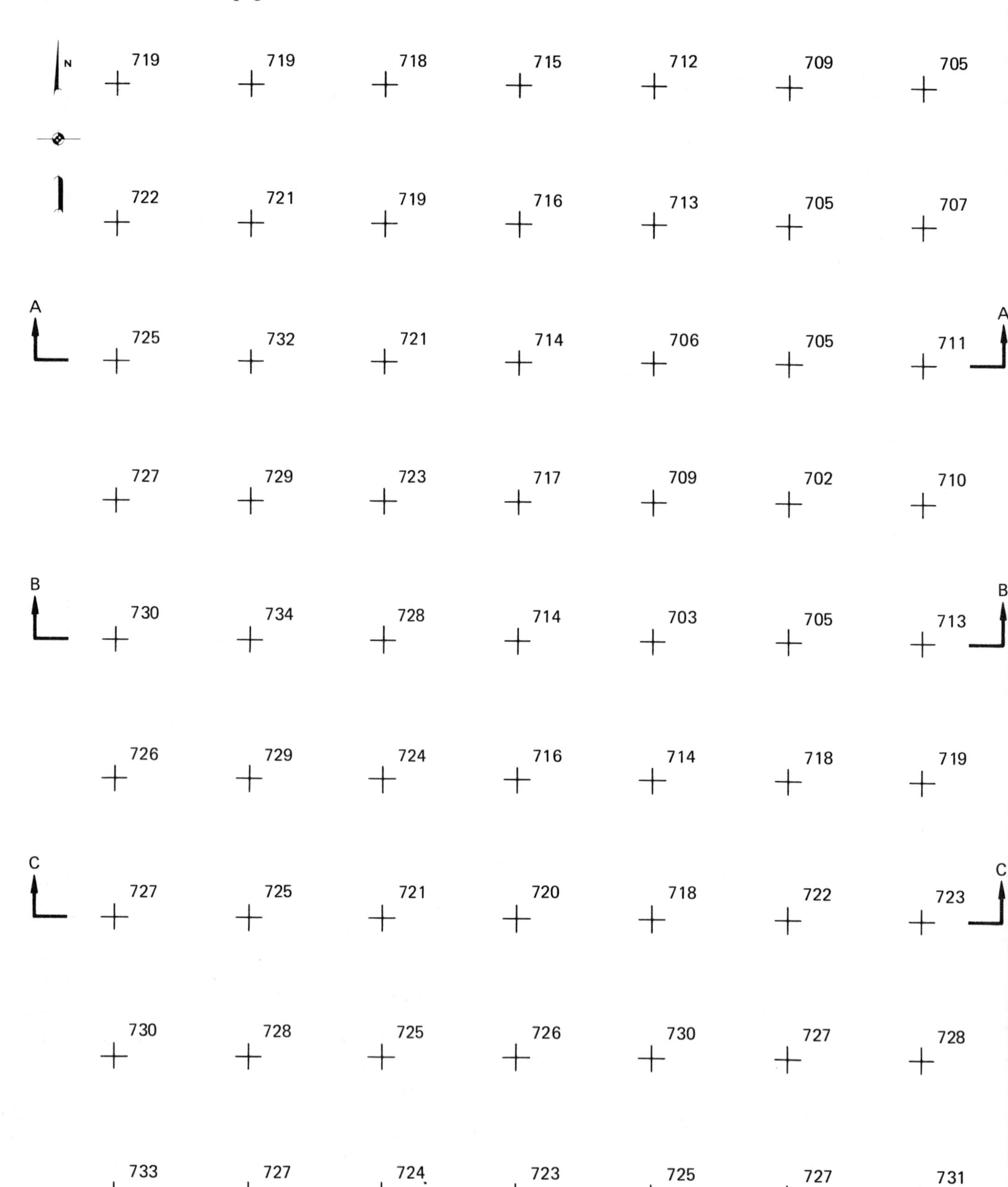

SCALE: 1" = 100.0′

HIGHEST ELEVATION: ____________

Practice Exercise 4-2

Using the answer from practice exercise 4-1, make the three profile views through A, B, and C. Shade in the ground area. Establish each 50-foot elevation. Show the pond at a 707-foot elevation. Compare your work to the answers on page 90.

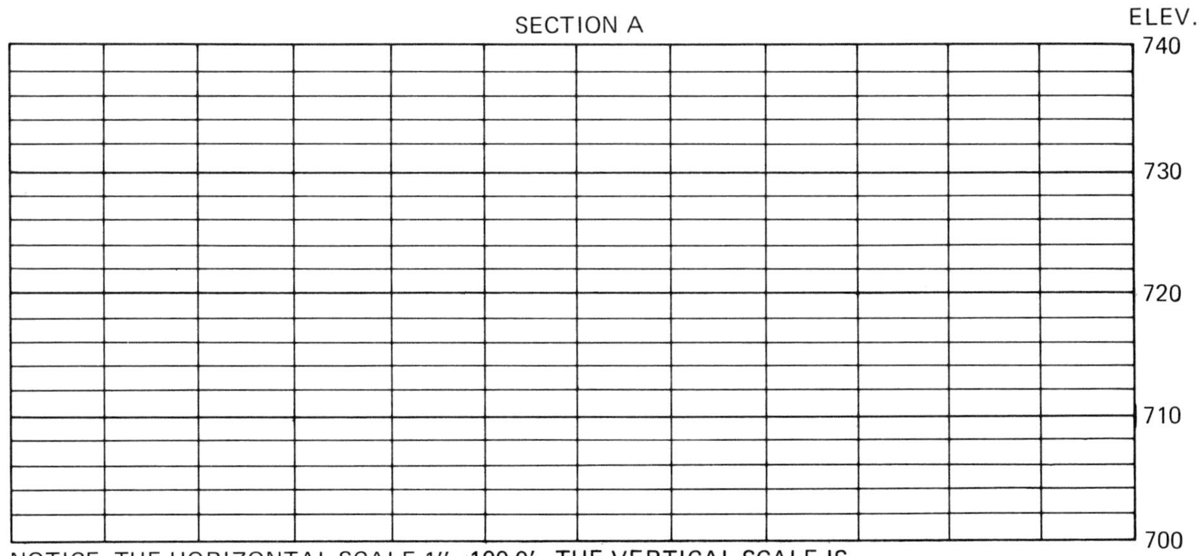

NOTICE, THE HORIZONTAL SCALE 1″= 100.0′. THE VERTICAL SCALE IS ENLARGED TO EXAGGERATE THE PROFILE FEATURES.

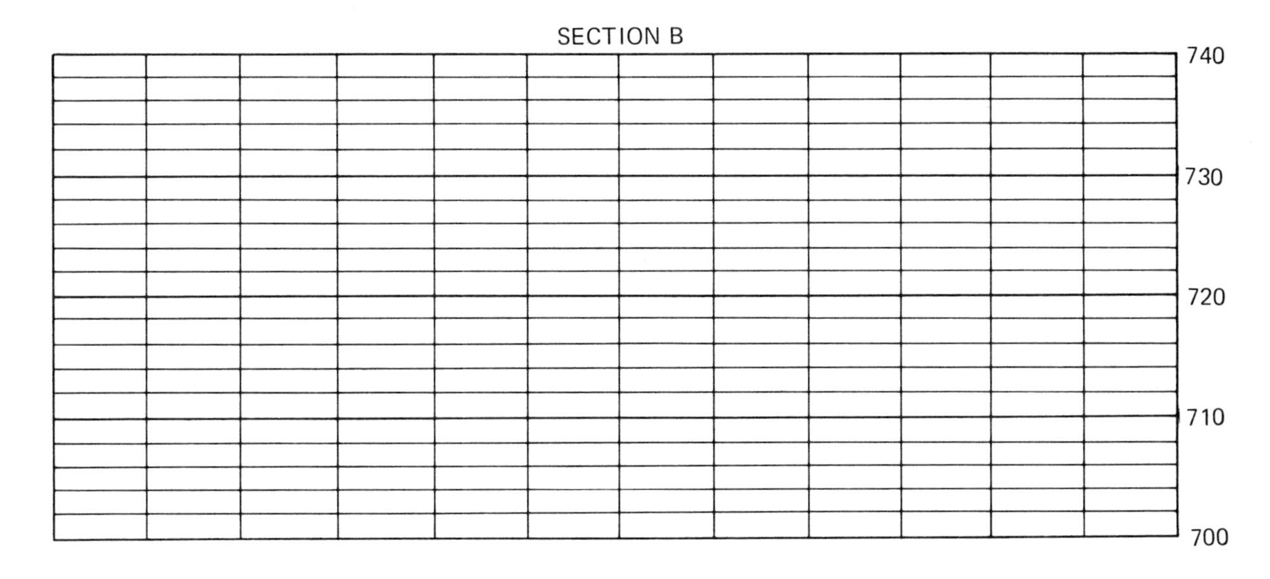

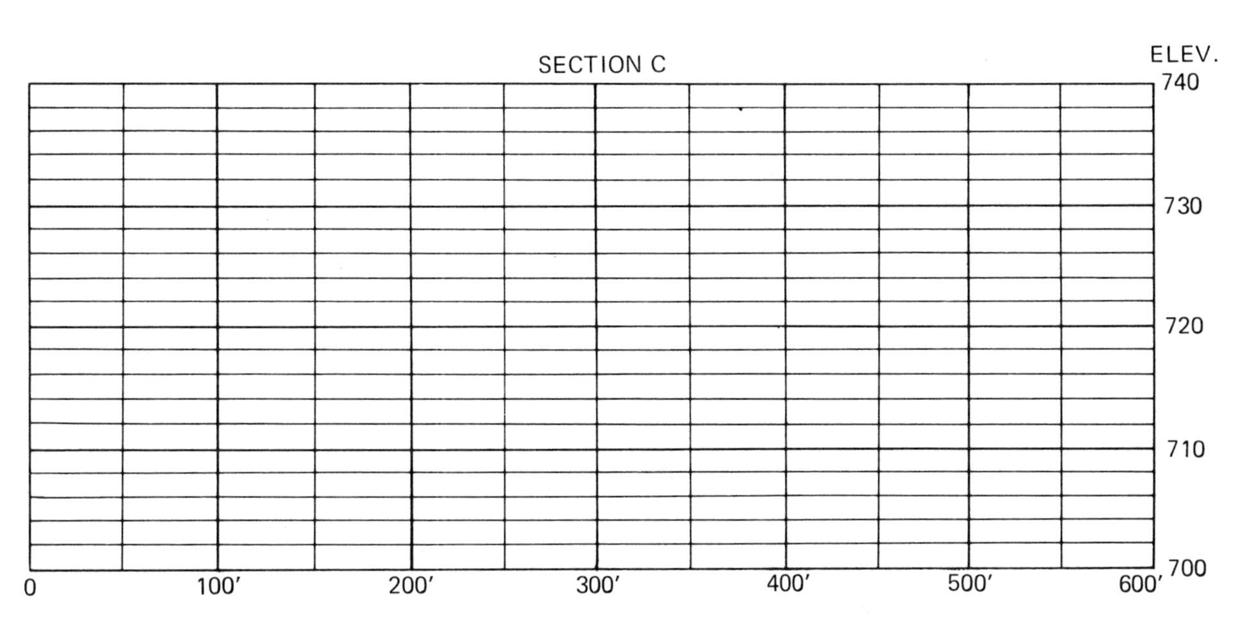

STATION POINT

Figure 4-11 is a sketch of a typical *station point* set in the field by a survey crew. A station point ususally has three stakes. One thin, long post holds the *colored flag.* The color is usually coded to mean center of the road, edge of road, bottom of bank, etc.

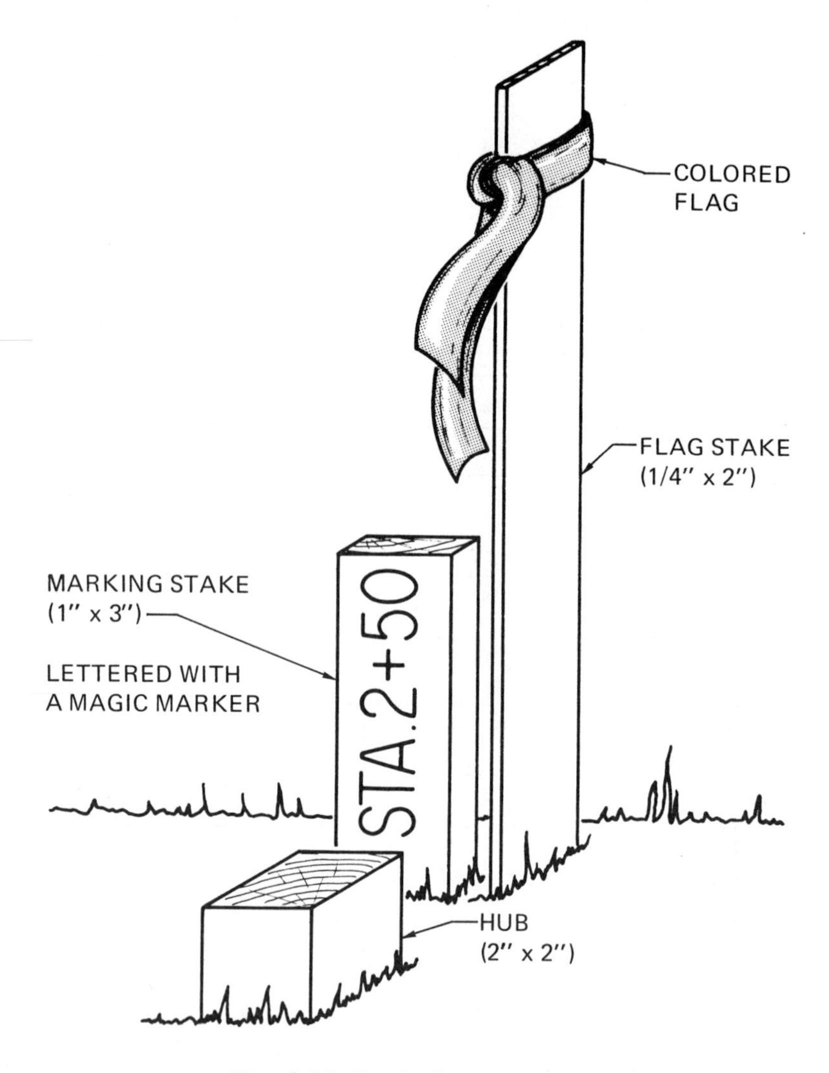

Fig. 4-11 Typical station point

Next to this post is a *marking stake.* A marking stake has the station number (STA) starting from the base or beginning point with 0+00. These numbers are in hundreds. The first number indicates hundreds from the starting point. 2+50 means 200 feet plus 50 feet, or 250 feet in all from the starting point. 5+37.5 would mean 537.5 feet from the base.

Next comes the *hub* which indicates line, distance, or elevation. It can give any combination of this information. Many times there is information on the back side of the marking stake, such as how much to *cut* or *fill.* "C" = cut and "F" = fill. C–10 means to cut down ten feet. F–4 means to fill in four feet from the top of the hub. Cut or fill is measured from the top of the hub.

Practice Exercise 4-3

Notice the existing grade in the drawing. A company is to build a road at the elevation of the new grade. Earth moving equipment must either build up the land (fill) or remove some earth (cut). It is the work of the drafter to tell them if they should cut or fill and how much. The first two answers have been completed. Finish the rest. Compare your work to the answer on page 91.

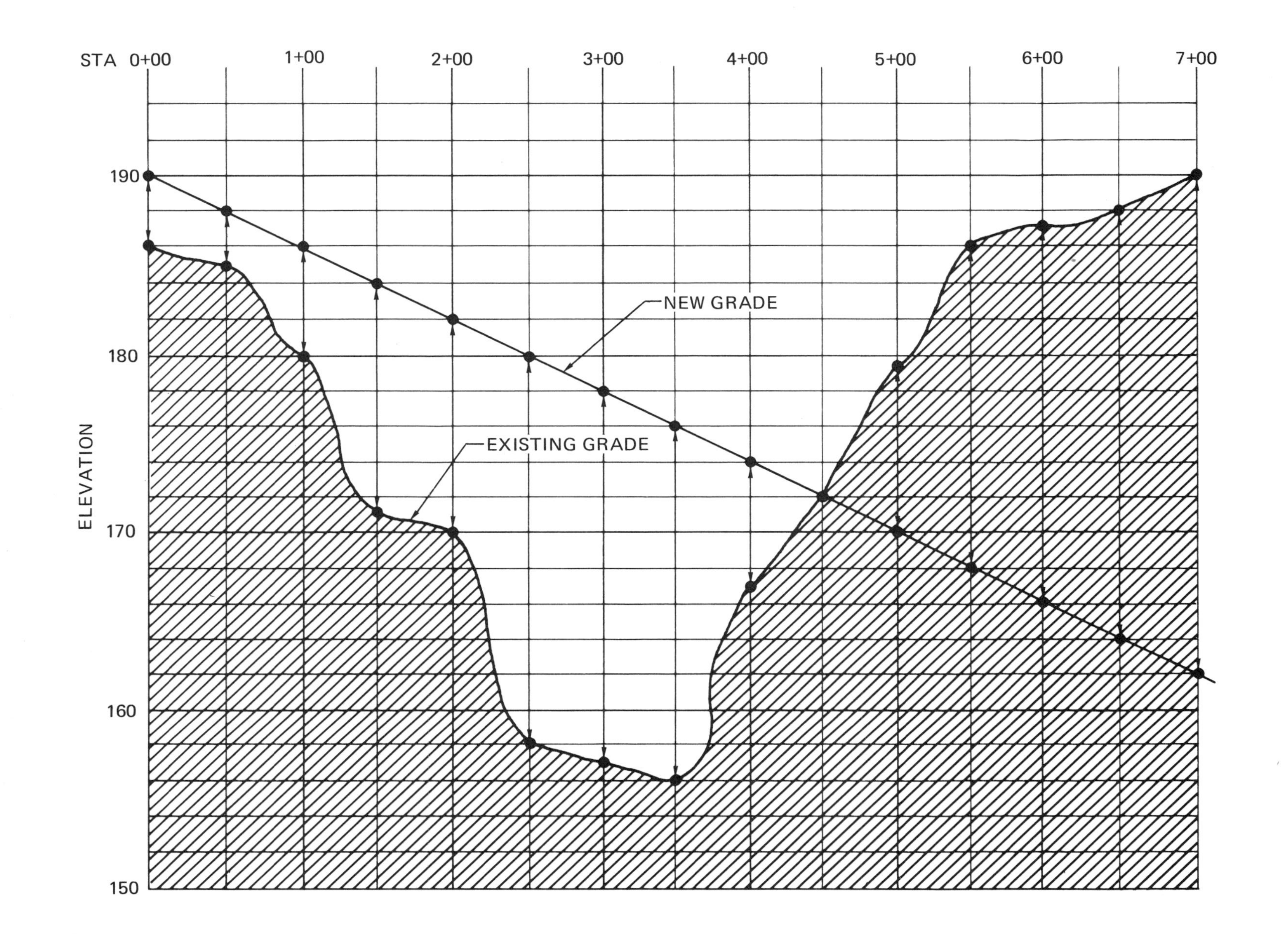
STA 0+00
1+00
2+00
3+00
4+00
5+00
6+00
7+00
190
180
170
160
150
ELEVATION
NEW GRADE
EXISTING GRADE

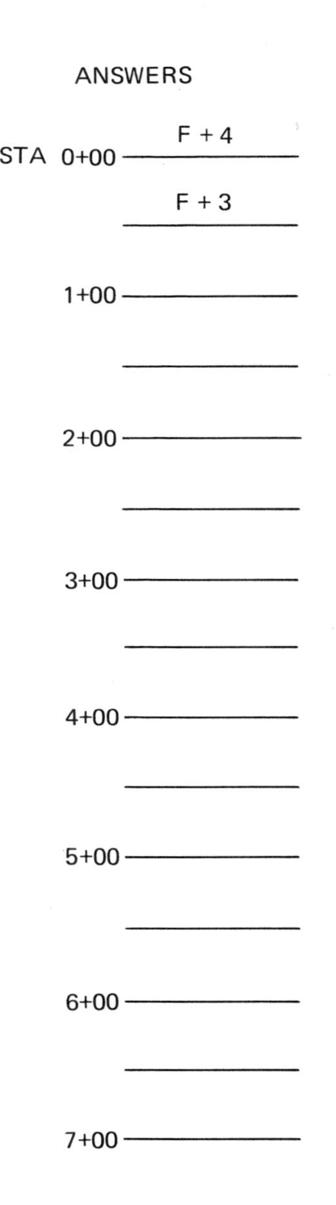
ANSWERS
F + 4
STA 0+00
F + 3
1+00
2+00
3+00
4+00
5+00
6+00
7+00

SLOPES AND RATES OF GRADES

Slopes, or rates of grade, are called off in three different ways:

1. Slopes by angle, figure 4-12 is used in mechanical drafting

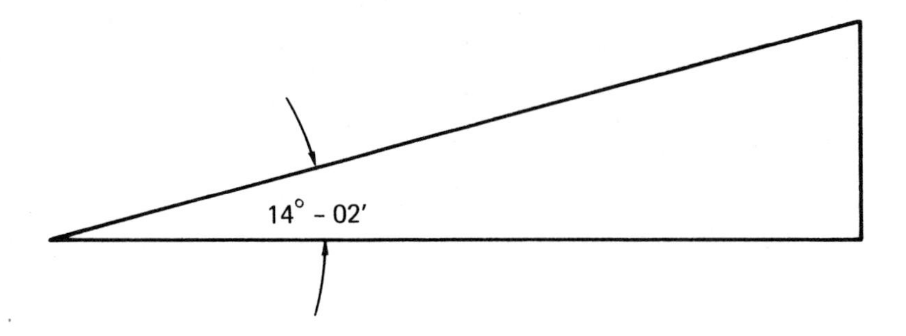

Fig. 4-12 Slope by angle

2. Slopes by horizontal to vertical, figure 4-13, is used in architectural drafting

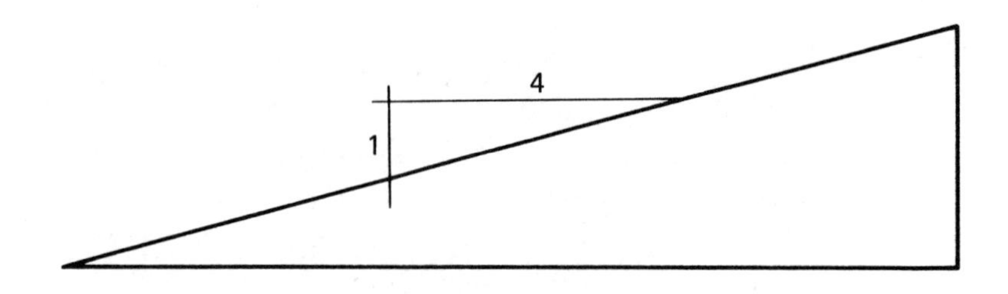

Fig. 4-13 Slopes by horizontal to vertical

3. Slope; by percentage, figure 4-14, is used in civil drafting. If in 100 feet the slope rose 25 feet, it would be called a 25% rise (+). Plus (+) indicates *rise,* while minus (–) indicates *fall.*

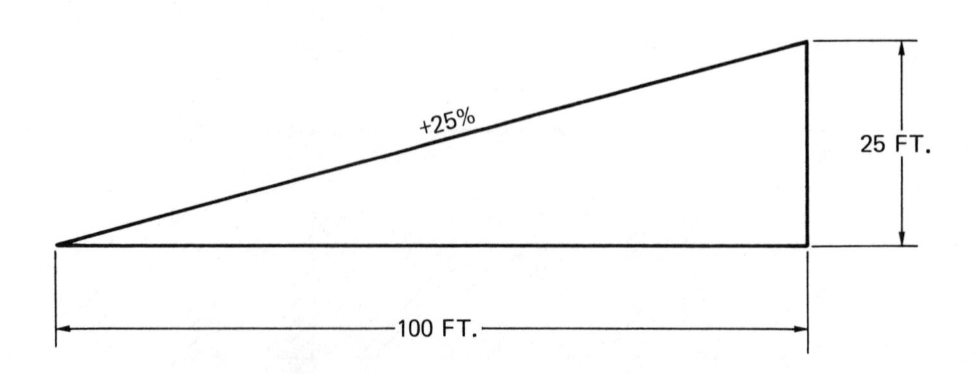

Fig. 4-14 Slopes by percentage

ANSWERS TO PRACTICE EXERCISES

Carefully compare your work to the answer for each exercise. Refer any questions to the instructor.

Exercise 4-1

Your answer should look exactly as illustrated. Note the various line thicknesses. Draw the exercise again if your answer does not agree within 99 percent.

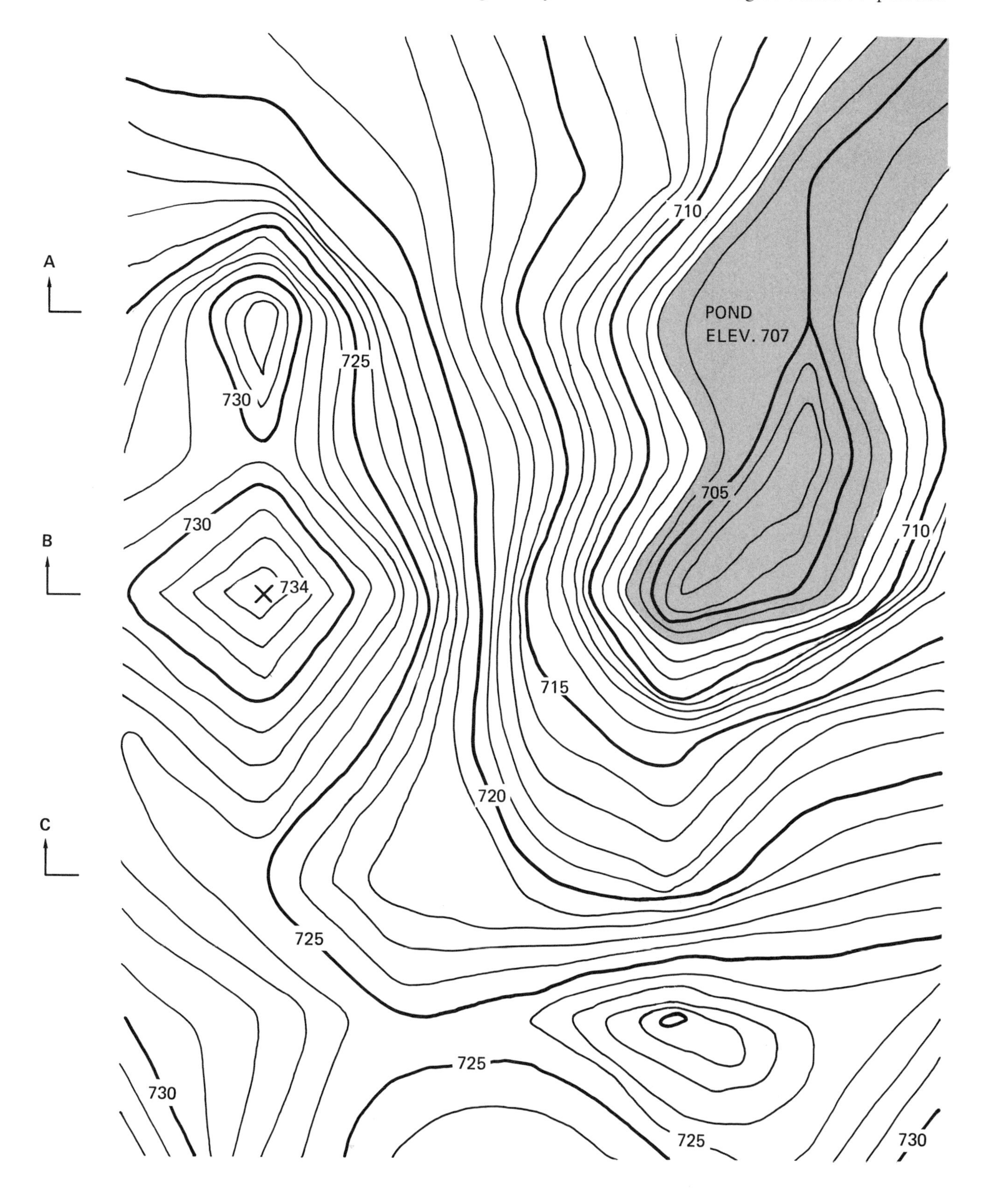

SCALE = 1" = 100.0'

HIGHEST ELEVATION 734
POND LEVEL 707

Exercise 4-2

Your answers should agree with those given. Note line thickness and water (pond) level.

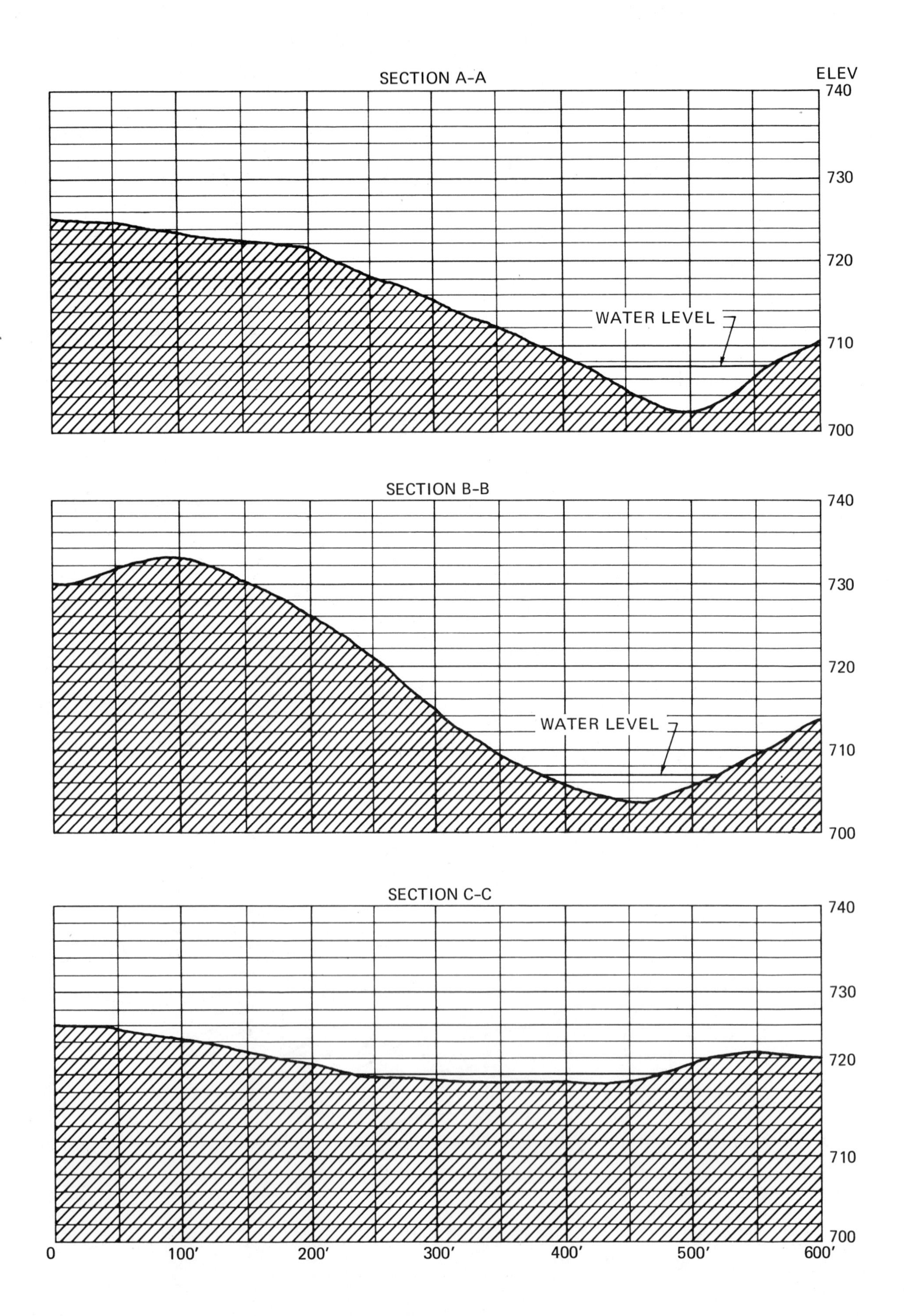

Exercise 4-3

The answers were calculated in even feet. The same method applies if tenths of a foot are used.

0+00 = F + 4	2+50 = F + 22	5+00 = C - 9
0+50 = F + 3	3+00 = F + 21	5+50 = C - 18
1+00 = F + 6	3+50 = F + 20	6+00 = C - 21
1+50 = F + 13	4+00 = F + 7	6+50 = C - 24
2+00 = F + 12	4+50 = F + 0	7+00 = C - 28

UNIT REVIEW

One-hour time limit

Using the field notes given, do the following:

1. Locate the center of a proposed straight road running from north to south. (The north end is located on the map.)
2. Assign this point: *"station point" 0+00* (north end located on the map).
3. Plot the existing grade on the graph provided and make a profile or section view of the land surface.
4. From point "A" on station point 0+00 (elevation 711 feet), draw a proposed road to station point 8+00 (south) at a +2 percent slope.
5. Add all notes and use correct line weight.
6. Add cuts or fills to the existing land profile required to bring the grade up to the new proposed road design.
7. Study the design. Are the cuts and fill totals about equal? Is it a good or poor design?
8. If it is a poor design, what could be done to improve upon the design and still maintain the required +2 percent slope or grade?

_______ Instructor's approval
_______ Progress plotted

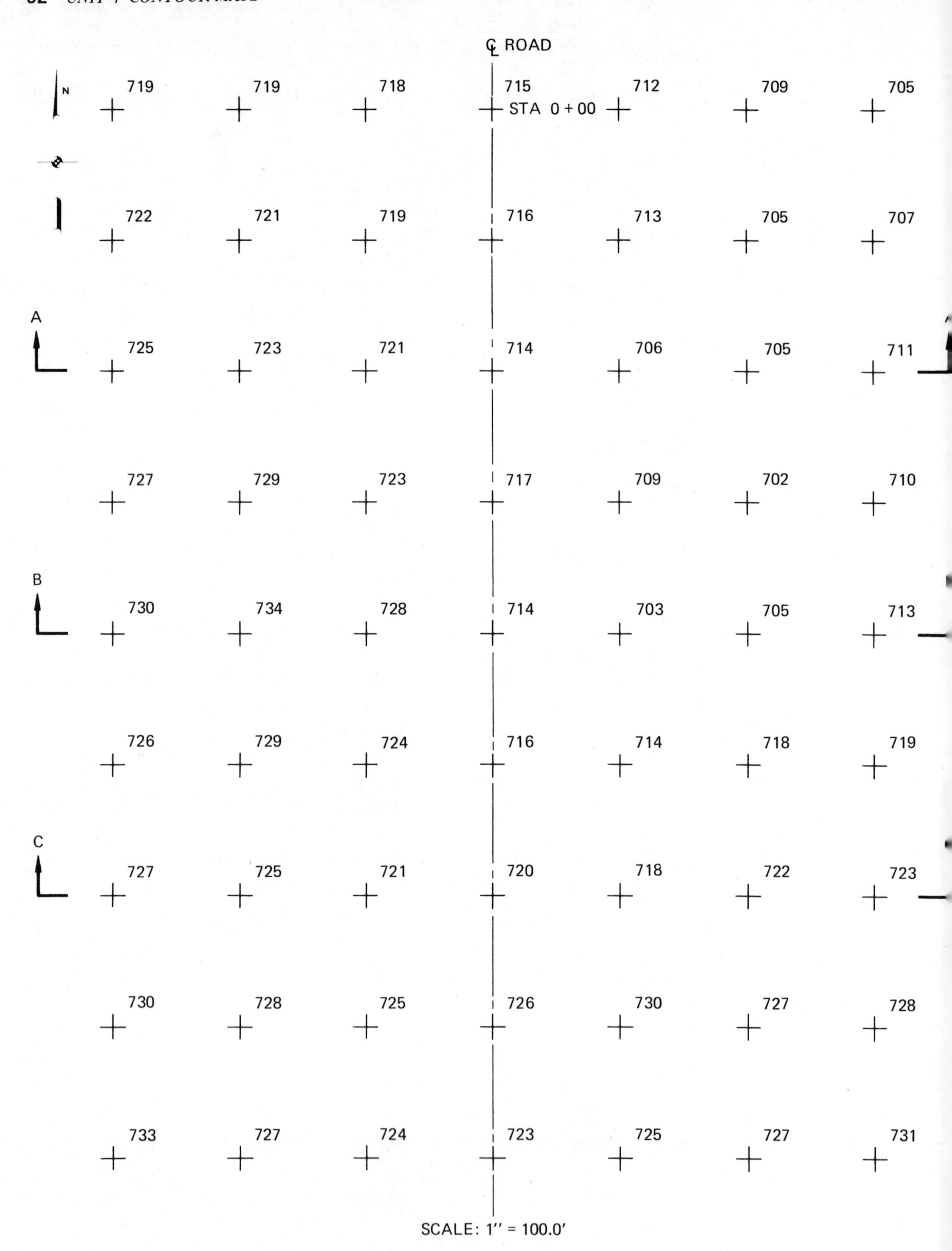

℄ ROAD
N
STA 0 + 00
A
B
C
719 719 718 715 712 709 705
722 721 719 716 713 705 707
725 723 721 714 706 705 711
727 729 723 717 709 702 710
730 734 728 714 703 705 713
726 729 724 716 714 718 719
727 725 721 720 718 722 723
730 728 725 726 730 727 728
733 727 724 723 725 727 731
SCALE: 1″ = 100.0′

STA.	C/F
0 + 00	
1 + 00	
2 + 00	
3 + 00	
4 + 00	
5 + 00	
6 + 00	
7 + 00	
8 + 00	

SOUTH

8 + 00

7 + 00

6 + 00

5 + 00

4 + 00

3 + 00

2 + 00

1 + 00

0 + 00

NORTH

730 725 720 715 A 710 705 700

ELEVATION

TOPOGRAPHIC MAP SYMBOLS

VARIATIONS WILL BE FOUND ON OLDER MAPS

Primary highway, hard surface
Secondary highway, hard surface
Light-duty road, hard or improved surface
Unimproved road
Road under construction, alinement known
Proposed road
Dual highway, dividing strip 25 feet or less
Dual highway, dividing strip exceeding 25 feet
Trail

Railroad: single track and multiple track
Railroads in juxtaposition
Narrow gage: single track and multiple track
Railroad in street and carline
Bridge: road and railroad
Drawbridge: road and railroad
Footbridge
Tunnel: road and railroad
Overpass and underpass
Small masonry or concrete dam
Dam with lock
Dam with road
Canal with lock

Buildings (dwelling, place of employment, etc.)
School, church, and cemetery — Cem
Buildings (barn, warehouse, etc.)
Power transmission line with located metal tower
Telephone line, pipeline, etc. (labeled as to type)
Wells other than water (labeled as to type) — Oil, Gas
Tanks: oil, water, etc. (labeled only if water) — Water
Located or landmark object; windmill
Open pit, mine, or quarry; prospect
Shaft and tunnel entrance

Horizontal and vertical control station:

Tablet, spirit level elevation — BM Δ 5653
Other recoverable mark, spirit level elevation — Δ 5455
Horizontal control station: tablet, vertical angle elevation — VABM Δ *9519*
Any recoverable mark, vertical angle or checked elevation — Δ*3775*
Vertical control station: tablet, spirit level elevation — BM × 957
Other recoverable mark, spirit level elevation — × 954
Spot elevation — × *7369* × *7369*
Water elevation — *670* *670*

Boundaries: National
State
County, parish, municipio
Civil township, precinct, town, barrio
Incorporated city, village, town, hamlet
Reservation, National or State
Small park, cemetery, airport, etc.
Land grant
Township or range line, United States land survey
Township or range line, approximate location
Section line, United States land survey
Section line, approximate location
Township line, not United States land survey
Section line, not United States land survey
Found corner: section and closing
Boundary monument: land grant and other
Fence or field line

Index contour
Intermediate contour
Supplementary contour
Depression contours
Fill
Cut
Levee
Levee with road
Mine dump
Wash
Tailings
Tailings pond
Shifting sand or dunes
Intricate surface
Sand area
Gravel beach

Perennial streams
Intermittent streams
Elevated aqueduct
Aqueduct tunnel
Water well and spring
Glacier
Small rapids
Small falls
Large rapids
Large falls
Intermittent lake
Dry lake bed
Foreshore flat
Rock or coral reef
Sounding, depth curve — *10*
Piling or dolphin
Exposed wreck
Sunken wreck
Rock, bare or awash; dangerous to navigation

Marsh (swamp)
Submerged marsh
Wooded marsh
Mangrove
Woods or brushwood
Orchard
Vineyard
Scrub
Land subject to controlled inundation
Urban area

ACKNOWLEDGMENTS

The author wishes to thank the following for reviewing the manuscript and providing critical input:

Clarence T. O'Brien, Land Surveyor
New York State Department of Environmental Conservation

Delmar Staff

Industrial Education Editor – Mark W. Huth
Associate Editor – Kathleen E. Beiswenger
Technical Editor – Harry A. Sturges

Illustrations

Engineering Graphics title page
U.S. Geological Survey 4-2
U.S. Department of the Interior Unit 3 pretest and unit review photos, 3-16, 3-19
Wild-Heerbrugg 3-18

Classroom Testing

The instructional material in this text was classroom tested in the vocational drafting department at St. Johnsbury Academy, St. Johnsbury, Vermont.